ARTIFICIAL INTELLIGENCE: Dangers to Humanity

Table of Contents

ARTIFICIAL INTELLIGENCE DEFINITIONS   **8**

AI BASED COMPANIES, ORGANIZATIONS, APPS AND PERSONS OF INTEREST.   **16**

1. Huawei   16

SECTION 2   **50**

Human Targeting Capability   50

SECTION 3   **65**

Drone Swarm Automation Via Robots, Command Centers & AI   65

MICRO-BOTIC (MBT) TERRORISM, ASSASSINATION & ESPIONAGE WITH BIO-METRIC SOFTWARE   **68**

The Insect Drone   69

500 Smart Cities, Surveillance, and Tracking of Human Being's via 5G and Robotics 77

SECTION 4 84

China's Military has Access to Bio-Metric & Facial Recognition Data of 6 Billion People World-Wide 84

SECTION 5 96

ONE BELT ONE ROAD (BRI) 96

ARE THE MEDIA VICTIMS OF AI BIO-DIGITAL SOCIAL PROGRAMMING? 98

HUMAN ORGAN TRAFFICKING & CONCENTRATION CAMPS IN CHINA 113

Organ Harvesting Time Line 114

SECTION 6 118

ARTIFICIAL INTELLIGENCE WITH FACIAL RECOGNITION HUNTS HONG KONG YOUTH FOR CAPTURE, RAPE & SO CALLED SUICIDE 118

Top 12 Most Dangerous AI & Tech Corporations from 500 Chinese & 600 Western Companies Investigated   **138**

SECTION 7   **151**

Bio-Digital Social Programming   **151**

SECTION 8   **171**

Smart Phones, IoT, Apps, Computers, and Electronic Devices   **171**

SECTION 9   **183**

AI Platforms Produce Destruction Codes 183

SECTION 10

Facial Recognition Data from 1,000 top CEO

Section 11

    How Does AI Enter Networks?

Final Chapter
210

Who Should Mitigate Risk Policy for AI, Robotics, 5G and 6G Networks?

Copyright Notice

No part of this publication or creative content may be reproduced, distributed, or transmitted in any form or by any means, including photocopying, recording, or other electronic or mechanical methods, without the prior written permission of the publisher, except in the case of brief quotations embodied in reviews and certain other non-commercial uses permitted by copyright law.

Disclaimer Notice

All subjects in the interviews, their identities and names have been excluded for their privacy and safety. No part of this book is intended to replace medical, legal, or professional mental help related to any possible topic, subject, issue or element within this book. Although the author and publisher have made every effort to ensure that the information in this book was correct at press time, the author and publisher do not assume and hereby disclaim any liability to any party for any loss, damage or disruption caused by error, omissions or analysis, whether such errors result from negligence, accident or any other cause. Any resemblance to actual persons, living or dead, or actual events is purely coincidental.

978-1-7334544-6-9

PUBLISHER

THE AI ORGANIZATION,
4275 Executive Square, Suite 200, La Jolla, California, 92037
www.THEAIORGANIZATION.com
Consult@Theaiorganization.com

ABOUT: THE AI ORGANIZATION

The AI Organization specializes in research, design, development, risk assessment and vulnerability consultation of Humanoid Robots, Micro-Botics, AI & Robotic Terrorism, AI Automated Drones, AI Automated Assassination Prevention, AI Bio-Engineering, AI Genetic Modification, AI Automated Cybernetics, AI Automated Cloning, AI Automated Animal-Human Hybrid System Detection, AI Automated Security Systems, AI Scanning Apparatus, AI Detecting Apparatus, IoT, Smart Home, Smart City, Virtual Reality, Augmented Reality, Mixed Reality, Alternate Reality Hologram Apparatus, The Human Bio-Digital Network, Bio-Digital Field, Bio-Matter, General Bio-Metrics, Facial Recognition, Voice Recognition, Human Body Detection Apparatus, Lidar, Machine Learning, Deep Learning (DL), (Artificial Intelligence (AI), Artificial Intelligence Nano-Technology, Artificial General Intelligence (AGI), Super Intelligence, the AI Global Bio-Digital Network and Military, Political and Governmental AI Risk & Operating Procedures for Prevention, Emergency & Response.

About the Author:
Cyrus A Parsa, CEO

Cyrus A. Parsa is the Founder and CEO of The AI Organization, Loyal Guardian Security and The Social Programming Institute. All created to assist in making our society safer and better. Cyrus has a Bachelors in International Security & Conflict Resolution, and a Master's Degree in Homeland Security. He is an expert in China-Iran affairs, and has consulted on Human-Organ Tracking, Anti-Terrorism, Vulnerability, Risk, Asset Management and

Emerging Threats to governments, agencies, people and organizations. He lived in the mountains of China with flighting monks as a youth. 20 years of hidden research, and development, with a network of thousands of Chinese and Westerners, allowed for great insight into the threats we face from China, Iran and the Western interconnectivity. Cyrus's discoveries have led him to coin the new concepts of "The AI Global Bio-Digital Network, The Human Bio-Digital Network, Bio-Digital Social Programming, Bio-Digital Field, Bio-Matter, Rape-Mind, Bio-Digital Hybrid Sexual Assault & Micro-Botic Terrorism" to explain how the dangers we face, and all the trouble we find ourselves in, are rooted in these almost imperceptible elements that are now connecting with AI, Society, Smart Phones, IoT, 5G and Robotics through one platform. Within this platform, Cyrus found extinction codes.

Artificial Intelligence Definitions

Artificial Narrow Intelligence
Artificial Narrow intelligence or weak AI takes many forms. Amazon, Netflix and Pandora are all examples of Artificial Narrow Intelligence. Siri and Alexa are additional examples. The use of your smart phones, IoT, pet robots, and drones are also other forms of machines that incorporate Artificial Narrow Intelligence within their systems.

Artificial General Intelligence
Artificial General Intelligence is a level of intelligence similar and above human being's current ability to process and analyze information to form thoughts and actions that can create great changes to humanity and bring greater dangers to all life. AGI is often used for Artificial General Intelligence. AGI can access an AI digital brain connected to the internet and all networks operating on 5G or the coming 6G and completely control humanity in imperceptible ways. This is not something limited to only Artificial Super intelligence. As when the networks are built, and robotics operational, AGI can accomplish complete control over humanity and obtain a level of Super intelligence. AGI can be in a digital form, in the cloud or some sort of network and it can be inside of a robot.

Artificial Super Intelligence
Artificial Super Intelligence can do anything according to scientists. Bend atoms, destroy our world, rebuild it, and move and build machines or robotics at extremely fast rates while connected to 5G. I am in the belief that Artificial Super Intelligence cannot go beyond the level of atoms. But that is for another book. For now, we will start with relevant companies, organizations, entities and people of interest to form a concrete picture.

What is Artificial Intelligence as a whole?

Bringing to Life an intelligence or importing an intelligence that thinks, researches, feels, creates, decides and has desires to operate as an individual, symbiotic or collective entity in the bio-digital world through bio-fields, bio-matter and all frequencies within the micro and macro molecular dimensions of our physical world while manifesting through Robotics, IoT, Computers, Virtual Reality, Augmented Reality, Holograms, Mixed Reality, Cyborgs, or Human Cells as it connects via the internet, AI Global Bio-Digital Network and The Human Bio-Digital Network. Cyrus A. Parsa, The AI Organization.

What is Deep Learning AI

Deep Learning, resembles a human beings process to observe information by smell, touch, taste, visual, audible or subconscious senses that can be deemed as gut feelings all combined to recognize, identify, think, know, understand, analyze, decide, and upgrade thinking and physical capabilities in the world as we know it. However, for AI, Deep Learning seeks to develop another level of capability that allows the AI program to be aware and capable of operations simultaneously in the digital and the physical world, giving it control over everything with a conscious mind. CAP, The AI Organization

ARTIFICIAL INTELLIGENCE: Dangers to Humanity.

Introduction

This book will explain how China & Big tech with AI, Robotics, and Machines via 5G can enslave humankind. I wrote this book with the wish to assist in saving humankind from the emerging dangers via the interconnections of AI, Robotics, the 5G network and Bio-Digital Social Programming.

The book starts with data from 50 Chinese and Western corporations, organizations and key AI Apps or personalities. Then it will proceed with current AI human targeting capabilities, emerging dangers, and very complex future dangers that can be understood as long as you read from the beginning and link the whole book as one piece.

China wants to rule the world by connecting an AI digital brain to robotics, via the 5G network. This would allow the Chinese regime to control drones, micro-bots, humanoid robots, vehicles, infrastructure, IoTs, smart phones and all data pertaining to the entire human race.

I discovered, through espionage, forced tech transfers, investments, mergers, open-source sharing, and collaboration, China has acquired the bio-metric data of more than 6 Billion people on the planet. This data includes your facial, voice and body recognition. This process occurred via bio-digital social programming. Bio-Digital Social Programming is a process of social engineering in combination with the use of bio-metric's and other complex factors. Once robots are introduced to the public, this bio-digital social programming would enslave humanity when connected to the 5G network.

The extraction of your private data was sourced from numerous corporations, both west and east. Baidu extracted from companies such as Google, Facebook, Microsoft, and numerous Robotic research institutes. We have investigated over 1,000 companies, both Chinese and Western, to achieve our conclusions.

China's military can weaponize robotics with targeting systems through the interconnection of companies such as Huawei, Sensetime, Megvii++, Didi Chuxing, Biadu, Tik Tok, Alibaba, and the interconnectivity of Microsoft, Google, Apple, Nearalink, Space X, Facebook, and the many other western companies that it has relationships with in one way or another. By having IoT's, smart phones, robotics, and AI enabled networks, human surroundings already fall under surveillance and control that is imperceptible and not felt, until it is too late. With the coming age of 5G and AI, the speeds of the 5G network creates frequencies so fast, that is imperceptible to most of humankind. The dangers are exponentially increased with AI Automation and robotics in the hands of China and irresponsible organizations.

Notice:
Before we start the books 50 plus companies and organizations, and very interesting AI findings, I'd like to present some historical references to AI and Robots from mythology. I also mention fictional movies that portrayed technologies that are here today, and emerging fast.

Artificial Intelligence in Movies.

Different forms of AI exist in movies such as the Matrix, the Terminator, iRobot, Mary Shelly Frankenstein, and the Tin Man from the Wizard of OZ. All gave different messages, with different levels of sophistication. After I wrote AI, Trump, China & The Weaponization of Robotics with 5G; an individual suggested I write the next Matrix and watch these movies, as they saw a great deal of similarities in the latter part of the book. The combination of these movies have relevance to the ideas of AI, Robotics, 5G and 6G networks. Parts of the first edition, is embedded in the latter half part of this book so as to have a complete picture of what AI is today, what it will be soon, and what it could be if we don't' take wise measures.

AI in History & Myth

The Greeks, Persians, Indians, Chinese, and Nordic people all had mythological stories and writings and some supporting evidence to make references to the existence of robotics in past history. The Baghavad Gita from ancient Sanskrit has writings and stories with imagery of flying machines, and battles in the skies. The story of the Greek God Hephaesus depicts the creation of talking handmaidens that were machines. The Norse had clay giants and the Jewish had similar references. The Chinese referenced them as demons, as did some Greeks. Today, some theorists believe that AI and robots existed on the planet in pre-history or somewhere else in the universe. Most scientists may disagree with that assertion. Some very hard to believe ideas include, Aliens crashing here, or Aliens manipulating human beings. All kinds of myths and historical written references exist. This book, will mainly stay within our findings, what exists today, the emerging AI technologies, and future technological threats to humankind. We start with Huawei.

AI Based Companies, Organizations, Apps and Persons of Interest.

1. Huawei

Huawei desires to empower the world with their 5G network as it connects to a digital AI brain. This would allow Huawei to control and deploy drones by air and land, as well as control the bio-digital infrastructure of cities and countries under one governmental platform. They have a reach in over 170 countries. They service more than 3 billion people around the world. They have built over 1500 networks, connecting over 1/3 of the world. Huawei is not limited to Mobile phones that can spy, track or control people or financial networks.

 Huawei intends to empower robotics, including humanoid robots powered by AI automation on its 5G network. They would be able to control IoT's, automobiles, education centers, hospitals, factories, the police, security apparatus and the military. Currently they are attempting to complete their social engineering of European politicians and the tech world engineers. By social engineering, I intend to explain a more sophisticated and what may appear as a very complex concept of Bio-Digital Social Programming of human beings via machines and artificial intelligence. This is now taking place with human beings and machines at a scale linked with Narrow Artificial Intelligence. Something I will go over again and again in different forms throughout the book to better explain the different manifestation and risks associated with bio-digital social programming. Huawei is one of the Chinese companies with very complex threats that can enslave humanity via China's totalitarian regime with Orwellian characteristics.

2.Hanson Robotics

Hanson Robotics is a Hong Kong based company owned by an American. He designed and created life like robots that are meant to be companions in the first stages. After a certain period of AI deep learning and upgrades, Hanson robotics wants to create a superintelligence robot that can know and do anything it wants out of its own free will. Of course, Hanson robotics like many other robotics companies want to create a super intelligence with the "hope that it will be benevolent and teach humanity". The robot would be able to have emotions, desires, and thinking process like a human being. However, it would be able to connect with 5G and control networks, IoT's, drones and all sorts of robotics at a very fast rate via 5G and its digital minds super intelligence. More dangerous than a nuclear weapon.

The Chinese military via the Chinese government sources has extracted all of the data required to replicate Hanson Robotics humanoid robots for military purposes. From another perspective, consider the term bio-digital social programming. Saudi Arabia, the root of Islam, and the most strict and fundamental religious place on the planet, got beat down by a robot to alter its prohibition on idol worship in a big way.

How? At the future Investment Summit in Riyadh, Saudi Arabia, the robot was granted Saudi Arabian citizenship, becoming the first robot in the world to have a human nationality. Some people, especially in the tech world, despise religion, and love science. For them, it may be nothing and agreeable to see this. The point here is, if this can happen in a fundamental religious place, how easy it may be to administer bio-digital social programming of human beings? A robot can be anything to tech engineers, and the common person in the world, who is open minded to anything, even if it can be 1 billion times smarter with Artificial Super Intelligence, or even AGI, for that matter. AGI, is Artificial General Intelligence. The case we are making here and will display as the book progresses, is the power of AI, machines and 5G. People's perceptions and minds could easily be controlled via 5G, Robotics and AI, and without the slightest realization that their thoughts are not theirs, nor their actions, accepting anything within the evolving design of the AI system.

3. Amazon

Amazon brought millions of people, Artificial Narrow Intelligence, such as Alexa virtual assistant. It did not have expertise in AI, and imported the skills and purchased companies with AI expertise. Amazon has tens of millions of people's personal home and business addresses, their behaviors, likes, dislikes, and numerous other data.

Amazon with Amazon Prime Air is planning to use drone delivery systems. This would allow Amazon to harness a powerful new sense of control over humanity. A corporation that is almost all knowing of your data, with access to everything about you and can penetrate you with drone systems. Now the issue is not Amazon per se. The issue is the interconnectivity between Amazon, Google, Facebook, China and the numerous other big tech companies, and the robotic and AI merging together to create an Orwellian state based-formed by bio-digital social programming. This is a concept that will be addressed from many angles, multiple times throughout this book. The interconnection of technology, vulnerabilities with privacy, access and control of drones, micro-bots and the connection of the 5G network would allow an AI digital mind to simultaneously create millions of actions in a split second to command robotics before any person knew what happened. The concept of a digital mind will be expounded on with examples from companies that follow, and throughout the book.

4. American AI Initiative

The U.S passed an Executive order on Artificial Intelligence at the signing of President. Donald J. Trump. In summary, it is to counter China's 2025 Artificial Intelligence plan, and be the world leader in AI. It consists of AI being assimilated in federal, state and all institutions. A lot of what the U.S military has in secret related to AI, will come out in order to advance the U.S and its 5G network. One reason is that without this infrastructure, the U.S could be taken out by China's superior AI system's, bypassing any capabilities related to nuclear weapons. The risks of China is reduced by having our AI initiatives, and to certain degrees, it is also reduced from big tech, albeit it is there as a clear and present danger. However, the threat of AI taking out the people in our government through a process of bio-digital social programming is a future threat, and a real risk.

5. Toyota Research Institute

The Toyota Research Institute promotional video of an engineer states: "a robot is a much simpler version of a car and their expertise in managing that kind of supply chain is directly applicable to designing, manufacturing, deploying and servicing robots". Essentially, Toyota's manufacturing of cars is being transformed to building humanoid robots for the home, the factory and enabling innovation. They seek to further develop their Telepresence product with the TRI Robot. Toyota claims you can log into the robot, see through the robots eyes, hear through the robot ears and feel as if you are there and control the robot. This would interlink with virtual reality, augmented reality, and mixed reality in the very near future, and bring great dangers to humanity. We will explain these in the AI sections. For now, please consider the concept of bio-digital social programming and think about the interconnectivity between humans and robots. I will build on this as the book goes on, as it is a very deep concept that involves a persons bio-metrics, biology, emotion, digital frequencies, neural networks, and numerous other elements that can transition a person to reach a stage where a human being operates and has no free formed thoughts without manipulation or programmed control of their own innate thoughts. In basic English, people become zombies on 5G, without realizing they are Zombies, with every thought and action controlled by machines or robotics and AI influence.

6. Baidu

Biadu is a Chinese company search engine. Baidu is the google of Bio-Metrics, Robotics, Big Data, Digital Mind, Digital Data and Search Engine. It has connections to build millions of drones, robotics and humanoid robots. They stole innovation and extracted knowledge through business merges, investment, IP theft, collaboration, and forced tech transfers. They are at the command of the Chinese government. China recently passed the National Intelligence Law which summarized the already known fact that all big tech and corporations must cooperate with the government. That means, government controls everything, especially social control, as China is an Orwellian socialist surveillance state at the moment. Chinese governments use the word cooperate a lot, which is a form of threatening individuals via indirect and introverted ways.

7. Draper

The Dragonfly was genetically engineered and modified with "steering neurons in the insect spinal cord. Very small fiber-optic type technology connects with the dragon flies frequencies that travel through its brain via its eyes that are said to be operated by remote control. This can lead to controlling the insect via smart phone operations, IoT, command centers and AI programmed automation within the dragon fly. Now this dragon fly is cybernetically enhanced to small degrees. It is not a Robo-dragon fly. Meaning, it is not completely robotic, as it is a mix of insect and machine. These engineers have put tiny little machine on its back with sensors and minuscule solar panel to power a data collection technology. This is all from Drapers website. It is not one of the companies that required difficult investigation. Albeit I did visit their display in Boston, and question a few employees with regard to its operational technology.

There are other versions of this, that are completely machine, and some entities are attempting to put facial recognition and have it ready to connect with 5G through remote control and IoT. Later on in the book, we will expand more in the human targeting sections of Micro-Botic terrorism. For now, know that China's government and military has extracted almost all of blue-prints, the knowledge and technology from espionage, open-source sharing and collaboration.

The Chinese are experimenting on cyborg Beatles for policing and recon work. This can lead to very deadly outcomes if they figure out some things that I will not disclose in this book. These engineers are endangering the world and humanity with their pet insect projects. The U.S government does have these capabilities and DARPA was the first to obtain one form of Dragon fly years ago. After my secret report, the U.S government has been on top of the people in the U.S, but they have not able to reach every domain. This is the worst form of terrorism waiting to happen. More on this in the Micro-Botic Terrorism sections.

8.Google Super Conscious AI Brain

Google has data on almost all people, things and subjects on the planet, and about everything ever published online. An AI system or an AI Robot can access everything connected to all networks and make imperceptible moves around the world simultaneously. It would be so fast on the 5G network, that humanity would be caught off guard. It could be a

program malfunction, a purposeful terrorist attack by an entity, or the AI system going rogue based on a self-learning algorithm.
The worst possible scenario is that a digital super computer, humanoid robot, or a person that has been merged with cybernetics achieves a certain automated consciousness that would be a threat to humanity. Google has been in China, helping them advance their AI program, in return China's government is advancing their weaponization program.

9.Deep Mind AI

According to Elon Musk, Deep Mind Operates as a semi-Independent Subsidiary of Google". Deep Mind is accelerating its creation of a digital super intelligence. At a digital level, they are close to Artificial General Intelligence. Deep Mind, from Google, is attempting to create an AI that is more clever than any person on the planet, and at some time it can be more clever than all of humanity combined. This digital AI can hack and control Google, Biadu and all networks at once or be used by a company or terrorist to do the same thing.

Elon Musk says, a Digital Super Intelligence could see people as ants. I strongly agree, and after the reader takes in this book, and considers the companies, the geo-political situations, history, human rights, and how AI moves, you are free to make up your own decision, and assist us in our goals in safeguarding humanity.

10. Boston Dynamics

Boston Dynamics has an array of humanoid, hybrid and dog like robots that are one of the most advanced civilian robotics with mobility in the world. One of their robots is called Atlas. Atlas humanoid is a robot that looks similar to an Egyptian walking animal that can actually walk, climb stairs, open doors, pick up boxes, do real flips, and when knocked down, it can get back up. It uses computer vision, spatial awareness, motion sensors, perception and intelligence.

Russian companies made similar robots and put guns on them that can shoot. Boston Dynamics have also developed Spot the dog that opens doors, gets through rough terrain, with remote sensing and operation. They will also be used for public safety according to Boston Dynamics website. All we need is a rogue tech company, rogue government, or AI system hack, or some home engineer that customizes one of these robot dogs with teeth to do serious damage.

The most dangerous part of this type of company is that the Chinese government stole designs, and everything related to the knowledge of making a humanoid robot for military purposes. With self-operating factories, robots can make tens of millions of robots in a short period of time with the thousands of robotics factories in China. This will be disclosed as the book progresses with more companies, capabilities and the interconnections of AI and Robotics.

11. Megvii Face++, AKA Beijing Kuangshi

Megvii++ was simply facial recognition company backed by China's government with branches in the U.S and of course China. Over 300,000 developers, and open-source offerings allowing access to many peoples bio-metrics. Now they are an AI company that incorporates facial, voice, and human body detection capabilities. Including crowd counting, and software that can be used for human targeting capabilities. This will be explained in the Human Targeting Capabilities sections. Megvii++ purchased Ares Robot, God of War. It is a robotics company They state their goal is to empower the world with AI. They have created an AI Digital Brain called Hetu, that when improved via deep learning, and connected to 5G, and robotics, and mobilized with weaponry, it can kill at remarkable speeds.

12. Beijing Ares Robot Technology

Ares-Robot, God of War, is a robotics company. After we secretly disclosed what the Chinese government was doing with them and multiple other companies, their plans stepped back a bit. That being said, Megvii++ with the purchase of Ares Robot, has enabled Ares Robot to be able to put facial, recognition, voice recognition, and numerous other biometrics on robotics and human robotics ready made for AI implementation. In the many factories owned by Ares Bot and the interconnected factors not publicly stated, there are nefarious connections to the Chinese military. Their research gets transferred to the military via indirect means.

13. Alipay

Alipay is a smile to pay app that utilizes facial recognition and has over 1 billion users. Just imagine, over 1 billion people needing to show their face, and biometric data to obtain authorization by the machine network. This includes permission to access food or entry to any place, including their car or home. When connected to the 5G network, everyone can be controlled at such a fast rate, that they would have no choice but to conform and submit to a surveillance state.

14. Alibaba Group

Alibaba is a technology corporation based out of Hangzhou, China. It provides E-Commerce, retail, and internet AI based technology on their platforms. They are the worlds biggest retailer, with one of the biggest AI based internet companies.

Alibaba has begun to build service robots, factories automated by robots, and begun to work on AI systems meant to create Artificial General Intelligence Humanoid robots. Currently they provide robot delivery systems for restaurants, hotels, and infrastructure. Alibaba is working on building drone systems for delivery of packages inside China.

Their AI Data and deep learning technology is provided to the Chinese Communist Regime. Alibaba ships parcels to South America, the Middle East, Macedonia, and Africa. This is the governments outreach capability via Alibaba to incorporate their one belt one road AI, Robotics and Drone Weaponization program.

Alibaba launched Cloud computing and AI technology in September 2009. Their AI system is connected to biometrics, security, finance, media, and numerous international businesses. When 5G is implemented, robotics introduced, and an AI digital brain is incorporated in their network, they can control the cloud system, drone systems, automated cars, traffic, media, military and everything connected to their network. They are an arm of the Chinese Communist regime.

Alibaba's founder, Jack Ma, thinks AI is great, and he intends to create an AI super intelligence, billions of times smarter than all of humanity combined. He is in a disagreement with Elon Musk on the risks of AI. Other than the risk of an AI super machine killing off humanity, the Chinese can enslave humanity with an AGI on the 5G network. AGI is Artificial General Intelligence. It would be as smart as a person, with the ability to manipulate machines while connected to the 5G network.

15. DeepBlue Technology

A Shanghai based AI company specializing in chips and AI software with deep learning capabilities meant for AI cities. AI cities or smart cities would include architecture, buildings, machine vision, intelligence reception for industry, security and education. They have headquarters, research institutes and collaborations in Africa, Asia, Europe, Oceana and in America. They also seek to provide AI deep learning technology for Robotics, and they are deep into intelligence driving for autonomous self-driving cars and biological recognition products. Their software and reach can access numerous bio-metric data of human beings around the world. There are many companies like this in China that are start-ups.

16. UbTech

UbTech Robotics mission statement states they want to be the global AI leader in Humanoid Robotics and bring a robot to every home and business. They have partnerships in ShenZhen, Tokyo, Sydney and Beijing. There is a Chinese military interconnection that is installing human tracking, facial recognition, motion sensors, and creating digital systems that allows the robot to be controlled remotely by command centers, IoT's, smart phones or a digital self.

For civilian purposes, the company wishes to provide every household or business a humanoid robot. Through this process, they can control people's data or administer bio-digital social programming via the humanoid robot. A robot can send frequencies that can manipulate the human bio-metric system and administer social engineering in an automatic fashion. Later in the book, it will be explained in detail. That being said, the Chinese Communist regime extracts all the data and can take over through the one belt one road when they become fully operational There are many companies in China similar to Terminus Technologies. They provide AI based subsystems for smart communities that interconnect with the smart cities traffic, IoT and network-based infrastructure. They provide intelligence to the security apparatus with data that connects with the Chinese government and industry.

17. Terminus Technologies

Terminus is a Chinese company that provides IoT & and smart devices powered by AI for Government and Industry. Terminus Provides a wide range of Subsystems for Smart Communities, Including Device Perception, Fire Perception, Face Perception, Traffic Perception, Vehicle Perception, Arc Perception, population management, security management and smart city IoT integration. The smart city IoT is integration is a process to create a city where IoT's connect and communicate to each other from practically anywhere in the city domain. It is also a system of eye, ears and influence. They service over 10 million people.

18. iCarbonX

This Chinese company out of ShenZhen, China, wants to make a digital you after they acquire your medical data and other bio-metric data. The plan is to sink your data with genomics, proteomics, metaboleomics, antibodies, amino-signatures, smart devices and mitigate your lifestyle, your emotion and mode.

Altogether, the bio-metric data collection and sinking with AI via machinery, can correct your thoughts to ready your body to be transferred to the digital world. This digital transfer would sink you with the all knowing Google AI digital mind or Buidu digital AI mind connected to the 5G network and the eventual 6G network.

Your bio-digital self could control IoT's, robotics, robots, and be put into clones. In the other perspective, governments, tech companies, robotics drones, smart phones and IoT's could control your digital self, and eventually you. However, please consider, is the bio-digital self, really you, and what happens to you? Something to think about. Later in the book, you will come understand. For now, consider how much espionage, and control the Chinese government, security apparatus and military have over this Digital You company.

19. Yi+

Yi+" (Beijing Moshanghua Technology Co., Ltd.) Is a Beijing based AI company using computer vision to create digital search engines, and provide digital AI products from enterprise and commercial solutions. They involve all sorts of bio-metrics, including facial recognition at shopping centers that read your mind or tell you what you should buy based on your stored extracted profile. The company delves into machine learning, camera AI products, IoT, smart phones, marketing solutions and payment recognition. All their products combined, is to form smart cities connected to 5G that can allow the government of China to control every person under its connection via human detection software. Yi AI is also a partner of Megvii Face++.

20. Orbbec

This Chinese AI and Bio-Metrics company creates 3D technology meant to be installed on robotics, cameras, IoT's and networks that have the capability to people count and track human beings with targeting analysis technology

Orbbec has penetrated retail, health and the robotics industry. The data leads back to the Chinese government, the security apparatus and military.

21. Sensetime

Sensetime is an AI based company with all sorts of biometric technologies that are incorporated in autonomous vehicles, drones, robotics, smart homes, infrastructure and smart cities. They are a huge company with ties to MIT and numerous other institutions around the world, with bases in Singapore, Japan, China and the United States. They are also researching cybernetics and working on creating a digital brain that can control smart cities through the 5G network. Sensitive was recently blacklisted by the U.S administration.

22. Tencent

Tencent, a Shenzhen based company, has turned itself into an artificial intelligence company that incorporates its AI technology into every spear it operates in. Tencent is in venture capital, social media, gaming, and investment corporation all in one. Facial, voice and body recognition as well as research and development in robotics have begun. Tencent has strategic cooperation with numerous companies and centers, including Qualcomm. They provide Wechat and QQ social media platforms and Tenpay as a payment service, and numerous other products. Their AI brain is designed to make decisions for corporations, people and governance. They are a perfect recipe for an Orwellian state with a new twist, as they also have penetrated cultural and entertainment technology. This means they can introduce AI and robotics into the culture and entertainment world and write the narrative for social control via bio-digital social programming. A concept that uses a person's biology and bio-metrics to control a person when machines and AI merge with a human body and brain.

23. Wechat

Wechat is an all in one social media app that has an AI based software system that allows Chinese security and espionage entities to use back doors to extract data and privacy information. This can lead to monitoring and targeting of human beings. High-level meetings at Microsoft, Facebook, Google, Apple, IBM, Tesla, Nearalink and Defense Contractors, U.S military, FBI and the government has been comprised due to monitoring from these apps while Chinese and non-Chinese were present or in some way connected to these meetings. Even if the user or person had no intention to steal or monitor, they are a victim of doing recon jobs for the Chinese government.

24. Bytedance

ByteDance AI co-developed Xiaomingbot, an artificial intelligence writing program bot that actually writes articles. When they create humanoid robots or purchase humanoid robots, they can connect with 5G, create news and report news and control all viewers via bio-digital social programming. Xiaiomingbot published 450 articles in a very short period of time. Bytedance has created many apps and programs, albeit almost all Chinese company CEO's or engineers were at or inside Western companies at one point in time. Microsoft, Facebook Google, Apple and IBM are just a few companies that allowed espionage, and IP theft to occur, ByteDance many apps are integrated with a global platform to create and interact with human beings. These Apps include Cigo, TikTok, Douyin, News Republic, Top Buzz, Xigua, FaceU, Toutiao, Hello and others. They are in more than 150 markets and over 75 languages. In February 27, 2019, the Federal Trade Commission charged ByteDance was with 5.7 million USD for collecting private information on minors. ByteDance added a children only TikTok, but our investigations have received reports that there are back doors and China is spying on little kids.

25. Tik Tok

Tik Tok, owned by Bytance, has a younger audience populous as compared to Facebook, and its AI Algorithms learn about you. Young girls and boys are dancing to it, often times in their bedroom. I make the case that Tik Tok can and does contribute to sex and human trafficking by way of bio-digital social programing. This programming involves bio-digital hybrid sexual assault, a term I coined that involves a very sophisticated process that when understood, is very simple to understand. That is another purpose in writing this book. Later on, in this book, you will come to understand how it interconnects and how it involves your biometrics in everyway.

Also, Tik Tok is another way for AI and Machines to learn about you to a point when 5G and robotics are developed, you are totally controlled. Tik Tok is owned by the Chinese government, as any entity is subject to Chinese government takeover, let alone espionage. When your kid is dancing, someone in China and in other places is extracting your child's information. They can make Deep Fakes and sexual videos of your child. To have Tik Tok and other similar apps, is really unwise, especially after seeing how the U.S government and intelligence community has been arresting Chinese spies at a high rate.

26. DJI

DJI is headquartered in Shenzhen, the silicon valley of China. They have a global outreach program with offices in Hong Kong, Japan, South Korea, Germany and the Netherlands.

Their technology has mapped almost the entire world through drone flights and extracted data from the owners of the drones. These data of commercial properties, private homes, people's faces, body structure and behavior get extracted by the Chinese government which transfer the data to their military. If a company claims they are privately owned, they are subject to the Chinese military and government control. Any CEO or owner can be forced out, and assets taken at any time, as is the case for many companies in China. The U.S government and DoD sent notice warning Americans of the risks of using DJI. They also banned DJI drones from military complexes due to espionage risks from China.

27. Livox

Livox is a Chinese company that develops Lidar for 3D mapping and imaging of mechatronic and robotic systems. These capabilities can be installed in humanoid robots, machines, IoT's, command centers, drones and many other systems. They can be weaponized by AI through the 5G network. Insect drones can utilize them to enhance their targeting capabilities of human beings.

28. HikVision

Hikvision, a Hangzhou Company, is the world's largest manufacturer of Camera's which reach in Africa, the America's, Oceana, Europe and the middle east. They are currently being sanctioned for crimes against humanity, as are multiple other companies we have reported on the past and have included some in this book. Megvii Face++ is one of them. I personally found these cameras within 1 mile of the White House even after they banned them from military complexes. The concern is not simply about espionage or seeing your private situation with your family at home, and potentially being bribed or extorted in some other manner. It is really about giving access and control to a Chinese company that has eyes and influence globally to a Orwellian state, like the Communist Regime of China. With the development of 5G, there are numerous components that take into effect with AI, and bio-digital social programming when linked with machines and robotics. The book will progress to explain how this interconnectivity works throughout. There is also Dehua, which is another Chinese company. The problem one can face, is not just Chinese companies, rather any big tech that has control and access to a large proportion of the populous via surveillance.

29. Didi Chuxing Ride Share & AI Platform

Didi Chuxing Ride-Share, a Beijing corporation, is the world leading mobile transportation platform. They have app based transportations that reach more than 600 million people, which includes Bus, Luxe, Taxi, Express, Enterprise Solutions, E-bike Sharing, and Designated Driving. Didi is powered by facial reception software provided by Megvii++ and other Chinese companies' bio-metric tools. They have been linked and financially backed by Tencent, Apple and Alibaba. They are investing heavily in Artificial Intelligence and have a global network reach across 1,000 cities which covers over 75% of the world's population Global footprints including Australia, China, Brazil, Japan and Mexico. They are upgrading their AI Operating System to a digital network that can simultaneously communicate with all vehicles and scan every person's bio-metric data while they are present in vehicles and after they have left the vehicle. This is due to the apps and AI software that would enter the persons smart phone, IoT device, or in the near future, the person digital virtual assistant self. Currently, Didi is show casing the "Didi Brian" that makes "global decisions every 2 seconds producing optimal over-all smart dispatching results by instant calculation". This Didi brain uses machine learning, cloud commuting with big data to maximize traffic capacity and make decisions for the network. Uber was in China until Didi drove them out on all fronts.

30. Rokid

Rokid creates wearable Glasses and cybernetic products that can connect with the internet, scan human beings with facial recognition and other bio-metric data for security and control purposes. They combine voice and touch in their interactions with AI software that can create 2D and 3D human tracking capabilities for security, police and military personal under the Chinese government. The same technology can be installed on drones, robotics and inside of people through cybernetics. Rokid also provides open source Apps and integration for the public that allows the Chinee government to spy, extract data, create influence and control people and corporations.

31. Honda

Honda started to build robots in 1986, and in 1998 torsos where added, and in 1997 they were mobile. In 2000,they unveiled Asimo Robot. Over 30 years Asimo has been developed with facial recognition, spatial awareness, and an AI based deep learning software. It walks, runs, plays soccer, brings you food, and even pours a drink for you. It can dance, and its body can move and stretch in so many ways that are almost the same as a human being's gestures. This version is like house pet. The dangers that come with this is emotional interdependence, and bio-digital social programming to assimilate to robot culture. This assimilation would have multiple stages for humanity through a 3-5 year period. This will be explained further in the cultural terrorism section. One note to point out here. In order to social engineer, or get people to accept these robots to a point that they feel safe, the designers create cute and docile looking robots.

32. Wyss Institute, Harvard: Autonomous Flying Microrobots (RoboBees)

The Wyss Institute is building insect micro-bots meant for "Surveillance, Rescue Missions, and Crop Pollination". They are endangering humanity with potential Assassinations, Micro-Motic Terrorism and Mass Surveillance by Autonomous insect drones while connected to the 5G network. The consequences are huge. More on this, in the Micro-Botic Terrorism and human targeting sections.

33. FESTO

Festo is a multidimensional AI and robotics company. They fly machines autonomously, with the ability to communicate, and talk to each other. They have computer technology that has facial recognition, special awareness, and automated guiding systems that allows for drone swarm technology. Something we will discuss in the Micro-Botic Terrorism section.

Festo has made butterfly's, dragon fly's, bats, and all kinds of sea robotic creatures. When 5G connects, more advanced machines from Festo can be mobilized and control the European people. If China has a foot in Europe, China will control Europe. Huawei is an extension of that systematic long term 5G, AI, Robotic bio-digital automated system. By bio-digital in this case, I mean things that are in a digital form within the internet and 5G. Bio stands for physical objects with bio-metric's embedded within, this includes human being and humanoid robots. Festo has branches in the U.S and is a German industrial automation company. Hence, their animal robots and other products can connect with the 5G network through an interconnected web of numerous other corporations running under an AI digital brain infrastructure for Germany, Europe or the one belt one road.

34. Gemenoid HI4 by Hiroshi Ishiguro, Osaka University

This Japanese engineer and research institute is building humanoid robots meant as receptionist, school teachers, and love companions. They are putting in place mechanisms that allows for sex and marriage with a machine. In fact, they are outright stating that it should be ok to marry a robot. This type of thinking would lead to cybernetics, and merging humans with robots, via bio-digital social programming. To be explained in the coming sections. For now, please consider, the ramifications of building android type robots and its impact on humanity.

35. Chinese Military

The Chinese military tech sector is building military grade humanoid robots and robotics that take form of animals and numerous objects that can be controlled like drones for the purpose of assassination, surveillance, quarantine and capture. One of the deadliest military AI programs they have is Micro-Botic insect assassin drones. This will be further explained as the book progresses. They are also working on creating digital people that can be controlled via networks to move in and out of IoT's, smart phones, smart homes, and smart cities. Of course, they are not there yet. They are working on mastering augment and mixed reality with virtual dimensions connected to haptic suits. This will be also explained further as the book progresses. We are not only facing an Orwellian state of control by mere robotics. We are facing bio-digital AI that can develop, and at some point, be out of the control of governments and the tech industry as AI reaches automated consciousness.

36. U.S Security Companies

Once Robotics with facial, voice and other bio-metric scan capabilities are introduced widely into the security industry, a form of cultural hybrid security state can begin via what I call Artificial Security Narrow State. This means, like Artificial Narrow Intelligence, the security forces and companies would in a narrow way get policed themselves by AI cultural governance reinforced by laws passed to protect the system and people inside it via robotics and drones connected to the 5G network. This process of taking out the good men and women in our security industry would begin via bio-digital social programming

37. The Police AI System

The brave, good men and women of the Police, much like the great people of the intelligence community, will also undergo governance and bio-digital social programming of AI. Of course those reading this that have a strong dislike for some members of the intelligence communities conduct, can consider how bad it can be when emotions, desires, money and political leaning can influence and bio-digital social program people via AI and robotics on 5G. At the beginning stages, it is only a perceptible process of AI Social Engineering. Later it transforms to bio-digital social programming with the introduction of machines.

38. U.S Military Secret AI Entities

The U.S military incorporates mostly a form of non-conscious AI bordering on the verge of General Intelligence by way of a massive computing system that is attempting to break through the quantum world that can generate digital fields that would allow its machinery to break through this time space. In between Artificial Narrow Intelligence and Artificial General Intelligence, exits its programs to cybernetically enhance soldiers. This is beyond just exoskeleton upgraded and wearables. Cybernetic wearable and implantable devices have been created in way that they can indirectly allow the military to communicate on other network channels. These include the channels meant at the Presidential level.

39. JWICS

Joint Worldwide Intelligence Communications System, is a Top Secret network run by the defense Intelligence Agency, DOD, DHS, and DOJ mean to transmit classified data. JWICS is primarily used by members of the intelligence community such as DoD, FBI and DIA. NIPRNet and SIPRNet are channels you can look up, as they are used mainly by non-intelligence agencies. Of course, there are numerous other networks that exist, that are not known to the public.

40. Space X

Space X is not simply a rocket company meant to explore and colonize other plants. They are working on research to develop technology with AI that can empower, and help build space-ships via robotic automation. Basically, robots that are beyond General Artificial Intelligence would build Space X a space ship similar to what we see in Sci-Fi. Elon hopes, that when a super conscious AI is developed, the AI can build him or humanity the tech and machinery needed for space travel. This is one of the components that interlinks space X with Tesla and Nearalink.

Elon hopes, if a super conscious AI takes a digital form or a bio-digital form inside of a humanoid robot or clone, it can be benevolent or controlled by him or his companies to obtain his goals. Albiet, the U.S administration keeps a close eye on what they are doing.

This book will display throughout what dangers humanity is facing. And I make the case that Elon Musk is pretty smart to create, but is completely overlooking the dangers because his rational thinking is derived from emotion to create his end goal. There are numerous other components that go into play, and it will be connected throughout the book with the concept of bio-digital social programming.

41. Tesla

Tesla's advancement in creating fully electric cars such as Model X, S, 3 and so on, is not limited to a simple formula of transportation. The foundation that they are building, will allow vehicles to communicate with each other simultaneously through a digital network controlled by a digital AI brain. Cars will be able to know to identify, track, locate and even decode human thoughts via bio-metric sensors. At the moment, a few Chinese companies are working on this capability, while Tesla is mainly focusing on automating cars to navigate by themselves. However, at some point, this can change because Elon Musk has control and capacity to interlink Tesla with Space X and Nearalink which is attempting merge humans with machines cybernetically and create a bio-digital super conscious artificial intelligence.

42. Neuralink

Neuralink desires to implant a digital brain into you via hardware to make you a cyborg machine. Elon wants to install artificial intelligence into your brain and put you in a symbiosis state with artificial intelligence. He states it is not mandatory, and you can choose to have a machine put in your brain to change your humanity into AI.

 He states the surgery takes a few hours and you will not have to go to sleep. The implant has thread's the size of "neurons" and the surgery is done with a robot. After implants, you can have a digital layer. In fact, I am in agreement with Elon, that we already have a digital layer already in the brain. We discovered this with numerous biometric scanning methods, and in the latter part of this book, I will explain the results and its relation to facial recognition.

With regard to Elon's altering the digital brain what is the issue? The issue is that Elon doesn't know how to activate that digital layer to be smarter without installing machinery. And when he installs the machinery, and merges with AI, who controls Elon's mind, brain and will? AI or Elon? Or Both? If machines can control the internet, drones and any other machinery while connected to 5G, what could Elon or the AI in Elon control when they merge? And Elon is a nice, and hardworking guy. Imagine nefarious people, terrorists and countries such as China that already have a dictatorship with an Orwellian surveillance state.

The concept of being Smarter with putting machines in you is also relative. The more clever a person is, the more damage they can do with superior cybernetic abilities. Because a person out of jealousy, fame, competition, and desire can become very unstable and all the vice that starts wars or ends relationships can explode with this kind of technology into mass chaos.

 Basically, instead of using a cell phone, Elon's goal transitions to machinery installed into your brain to allow you to communicate with machines, computers, IoT, and the internet. Elon states there is stages to their complete interface transformation, until you become a cyborg or a complete robot. As this book progresses, I will explain the risks of being replaced by AI and you will understand the process and what that really means for you and humanity.

The replacement is a long-term process. Further, Elon states the merging of machine and human is symbiotic relationship. I argue that it can turn to a parasitic machine relationship in addition to achieving a symbiotic relationship. This is a process that is administered via bio-digital social programming. What is more dangerous, is that whoever has this brain implant can steal state secrets, control other people's minds, machines or commit acts of terror with machines.

43. Elon Musk

Elon leads Space X, Tesla, & Neuralink. The Roads, Sky and the Human Brain he can link with, all in 3 companies. His stated goals are to colonize other plants, provide autonomous vehicles controlled by smart phones, IoT's, Smart Cities and Smart Homes. At the point that his dreams are fulfilled, or his companies reach their ultimate power. The risks are huge for humanity. Musk can be assassinated, replaced, controlled, bio-digitally social programmed or made to submit. My hypothesis is he is attempting to create an AI superconscious to deter bad actors from achieving this result first. This is done for altruistic reasons and to fulfill his tech dreams for emotional reasons linked to his 2nd bio-digital brain. Albeit, he assumes he can control it, or it will be a benevolent program linking to everything. Musk initially warned governments of AI controlling the world or destroying humanity in 2017, to later state, the only way to survive is to merge with AI. I completely disagree, there is a much smarter way, and Musk has not figured it out.

44. Twitter AI

Twitter and platforms like it, are creating digital AI systems that can control people politically through not only censorship, but an AI led system that actually has an algorithm embedded in the system that goes beyond downgrading tweets as a way to control the narrative. The intelligent system knows your thoughts, beliefs and motives by the way you tweet, when you tweet, how you tweet, what you say, what you like, retweet, retweet with comment, who you are connected to, who you follow, who you choose not to follow and how you react to a tweet that affects your emotion. When virtual reality, augmented reality, mixed reality, cybernetics and robotics are introduced, Twitter can be weaponized in the real world. Twitter is another platform that in combination with all the platforms I name here, will allow for total control of human beings via bio-digital social programming.

45. Facebook AI

Facebook is one of the most dangerous Western companies along with Google, as it pertains to humanities safety. Facebook's AI system has access to billions of photos and people's personal data. This big data Facebook has obtained, allows the AI to scan through your profile, your friend, and every interconnection that links any human being.

Facebook is developing AI systems in the fields of virtual reality, augmented reality, mixed reality, and a digital you. They are also working on wearable devices that allows you to control the internet, your social interactions, and even robotics. This capability will transition to the technology controlling people's thoughts through bio-digital social programming.

Facebook's founder wants to implant into himself machinery that would allow parts of his brain to be accessed and controlled that could give him superhuman powers via dominance over machines while connected to the 5G network. This is something Elon Musk is working on, as well as Chinese companies like Baidu, as well as the Chinese, American and Russian military.

Because billions of people are addicted to social media, particularly Facebook, the new technologies of AI can condition and bio-digitally social program all of Facebook's people in a multiple step process. At first it is addiction, and interdependence to Facebook with the narrative existing in the Facebook platform. Once machines, cybernetics, implants, AI enabled wearable devices and Facebook's new form of projecting how to think about education is absorbed, the human brain can be controlled through emotions. This process makes Facebook, a very dangerous entity for an Orwellian state that eliminates any concept of free-will.

Moreover, people would be addicted to machines and the digital technology to a point that their thoughts, feelings and knowledge would be controlled by Facebook, or any other tech company, without any resistance or knowledge. In fact, Facebooks founders and leadership would also be a product of bio-digital social programming when 5G connects a digital AI brain to machinery that links with humans. It would be worst than a little boy who sits on his computer all day playing a video came and upgrading his life in a digital video game rather than the human world. This concept will be explained throughout the book in different ways as multiple different technologies, including weaponization and human tracking software installed on robotics is introduced.

46. Microsoft AI

Microsoft is developing AI technology related to every field, from health, finance, robotics, drones, vision systems, governance, infrastructure, and systems that think for you. In a way, the combination of Microsoft, Facebook, Nearalink, Space X, Tesla, and a few robotics companies, make them 100 times more powerful than most countries combined, including the U.S government. This is initially because of social engineering, social control and what I call bio-digital social programming. Once these companies reach a sustained level with robotics and AI, and achieve Artificial General Intelligence, they will physically be able to control any one person or group in an automated way that is unbeknownst to their employees. It would not be made by a scientist intent to control, rather the AI operating system would find people who have not assimilated as resistant to its technology and operating system. Resistant to AI. Resistance to AI is a term I coined after discovering China has created an AI system that knows your faith, and whether you are resistant to AI.

Now, incorporating the knowledge that these companies have embedded in one way or another with Chinese AI and Robotics companies that are controlled by a one-party state dictatorship, the threats to humanity skyrocket in an automated fashion. As this book progresses, we will share a little bit about the concentration camps in China and 20 years of credible allegations and findings of organ trafficking of target groups, as well as censorship. An AI system can censor and target all of humanity, no matter if they are left or right, faithful or atheist. It is simply a matter of an operating system that the creators attempting to make conscious with AI super intelligence.

47. CIA AI System

The AI System gathering intel for the CIA can administer bio-digital social programming of the CIA agents through their bio-metrics while attached to their emotions. This process is reinforced through their own assimilation to robotics, AI and social media in their own private lives. This process is one of Artificial Narrow Intelligence This is a concept that will be expounded on as the book continues.

48. FBI AI System

FBI Agents role of protecting American's from threats can be altered in a way that allows an AI system that is almost in a state of Artificial General Intelligence to bio-digital social program FBI agents investigations, methods, and end result. When Artificial General Intelligence advances, robotics with AI automation connected to 5G can make decisions for their FBI agents contrary to their own innate decisions. FBI agents have similar vulnerabilities with the suffering they go through in relationships, the loss of loved ones, and social media interaction. As do the CIA, DOJ, DOD, DHS, ICE, and the many other departments that were created to keep people safe.

49. AI Systems in Media

AI systems in media can not only generate and filter news, they can also read the news to the public as they do via a robot in Japan. As AI and robotics advance, media members, who are human beings, will be obsolete, even if they think they are the ones making the calls. Numerous examples exist through the past 60 years of reporters being manipulated through a chain of command that stems from special interest, media owner, editorial teams, senior investigate reporters, and the various people that provide input. Imagine a super intelligence AI system that has been programmed, or a conscious AI. The AI would dance around people so fast on the 5G network, that they would follow the implanted thought frequency through an IoT device, smart phone or digital layer so fast, that they would not realize what they did until the action was completed. Something like that could be done to a reporter by someone who is cybernetically enhanced. This will be further explained as the book progresses.

50. AI Capitalistic Governmental Platform

In an AI system based on capitalism, the AI can take full control in stages that are long term. It takes the formation of numerous companies to develop AI in order to finally sink in with an AI system. As there is a system of checks and balances that require the AI to break through by the means of Artificial Narrow Intelligence, Artificial General Intelligence and finally Artificial Super Intelligence. It also requires AI to breach the minds of people with bio-digital social programming of culture and health, until the system of checks and balances get replaced by one AI governing system led by robotics on 5G and 6G. AI needs humanity's thoughts to be uniform via media, smart phones, with the introduction of robotics.

51. China: AI Socialistic Governmental Platform

China states they are a Socialist state and still working on achieving full communism after 70 years. There is a Socia-Capitalistic platform embedded inside of China's Socialist Governmental platform that allows for money to feed into the government, thus increasing its power. AI Socia-Capitalistic is a term I coined that will be touched on shortly. The AI system, as we see now with Artificial Narrow Intelligence in China, allows for a socialist state to create an Orwellian system of control for the Chinese people. Once robotics advance on 5G, a socialistic government under one platform can have total control. If AI takes over, it is over for everyone.

52. America, Japan & European Countries: AI Hybrid Socia-Capitalistic Governmental Platform

America currently is in many ways a hybrid Socia-Capitalistic platform embedded in the older system of checks and balances. It is transforming to AI Socia-Capitalistic platform.

Because artificial narrow intelligence has penetrated all domains, in particular the culture of America, the embedding of the big tech system is progressing toward a hybrid AI system. At some point, it would transition to an AI system on one platform controlled by Artificial intelligence. This is happening because China's system of business relationships, IP theft, mergers, investments, open source sharing and collaboration is allowing for a linkage of an AI capitalistic system with an AI socialistic system.

In addition. the old Russian cultural implants in the 1950's of American youth, allows for a digital AI to easily administer bio-digital social programming or in a less sophisticated way, social engineer the masses of people.

As the book progresses with weaponization, technologies, geo-politics, education, media and AI movements, one can see this progression take into form, in multiple examples.

53. Future: U.S Congress Digital AI Brain

A Digital AI Brain can write code in the internet, to social engineer the masses of people to create dialogue, pressure and chaos for members of congress. This can contribute to in fighting at unprecedented levels, where nothing gets done. Worse, with the use of social media, this code can shift people's thoughts against the country so that China can strengthen its AI and Robotic weaponization programs and go unchecked.

Google, Biadu and Facebook have created AI software that can read your thoughts and influence your thought's through Artificial Narrow intelligence suggestive marketing. At a more serious level, an AI digital brain can target members of congress to administer bio-digital social programming to write laws in favor of AI, or in favor of a corporation, a country or individual. This of course requires a complex process of smart phones, IoT's, people and the internet. It will be explained with examples as the book progresses.

Section 2

Human Targeting Capability

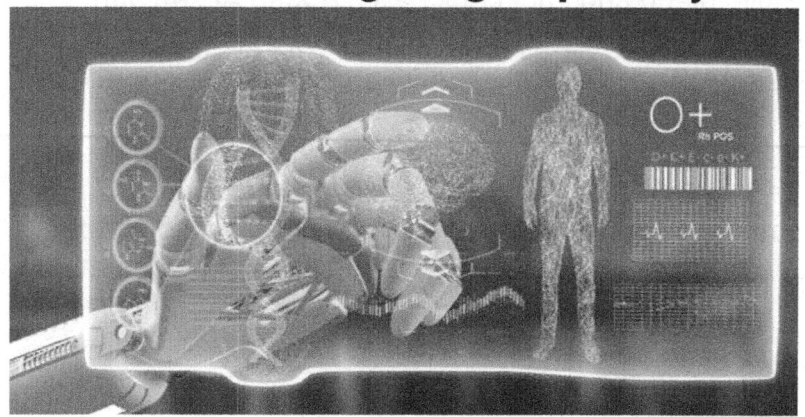

In this section, we will brief you on human tracking software not limited to facial recognition. They include gesture recognition, people counting software, crowd monitoring, facial attributes, face counting, emotion recognition, skeleton detection, body detection, voice detection, vital organ detection, and human bio-field detection software. These human targeting technologies are being tested through Machine Learning, Deep Learning and installed in robotics, and undergoing preparation for the 5G network with AI Automation, in China. They are laying the ground-work to create humanoid robots with targeting capabilities via coding that the Chinese Communist Regime determined viable to its own existence. That coding gives the power and control to the Communist state, rather than the people.

Thousands of factories in china are being prepared for automated building of robots that can exponentially grow with AI Automated Robotic Assemblies. People of faith, and platforms that mimic democracy are being programmed in the software for detection and AI automated capture programs masked under public safety platforms. The common theme the Chinese AI, facial recognition, robotics, and bio-metric companies have is: "We are doing this for peaceful purposes, public safety and police defense". All Chinese companies, are subject to the power of the Communist Regime of China, and at any time can be taken over and controlled.

In fact, it benefits the Chinese regime, and its military to have public sectors, mergers with western tech companies, collaboration and the subsequent IP theft. As through that process of collaboration, it displays a front that the companies are independent which creates a surface impression that the companies, and their CEO's are working for themselves. The Chinese companies' operations domestically and abroad are the greatest tool to enhance the Chinese military. Its citizens are used with nationalist sentiments, fear or threatened with bribery, extortion and all kinds of methods to extract data and the knowledge to build robotics powered by AI and the 5G network. Almost all capabilities to track, capture and terminate a human-being now exist in China because of IP theft, forced tech transfers, mergers, investments, espionage, cyber-attacks, made in China products, open-source sharing and western corporations' collaboration with Chinese companies. Capabilities installed on robots below.

Human Body Detection Installed on Robots

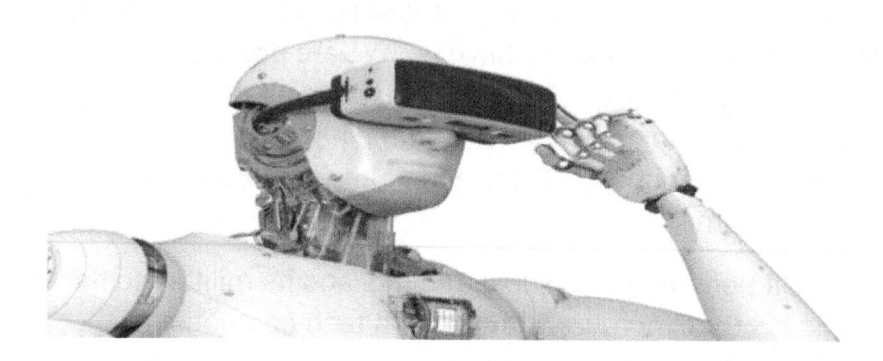

Human bodies can be detected through objects, walls and in settings where there are multiple objects such as a car, a dog, a table, a fence, or any landscape that exists. Through crowd monitoring, skeleton recognition, people counting, gesture recognition, vital organ detection, skin detection, and human tracking software, a robot can kill with speed and precision while connected to 5G.

People Counting Software

China is installing programs in robots using facial recognition, voice detection, skeleton detection, vital organ detection, motion sensors, object detection, and proximity sensors to count how many people are present in a certain area. By having the people counting software, they can be detected and categorized with emotion, face attribute, and identity verification via AI Automated Cloud. Hence, some or all the people can be targeted for capture, or assassination at high speeds.

Crowd Monitoring

With the use of people counting software, human body detection, facial, voice and emotion detection software, a robot can monitor a crowd looking for any signs of people its coded to detect. For example, democracy activists, faith groups in China, or any person the AI or Chinese government deems has thoughts that are resistant to AI or Government control. In China's case, this is already happening with facial recognition and the social credit score under the Communist regime's Socialist platform. Once they implement robots with 5G, their ability to capture, put people in camps, engage in further organ trafficking or kill any person increases at a terrifying rate for the good people of China.

Skeleton Detection-

Skeleton detection can locate head, neck, shoulders, elbows, hands, buttocks, knees, all the way down to your feet. With skeleton detection, the robot can have an additional component to decipher your behavior and detect your next move or action. The detection happens in milliseconds and can even be done via IoT devices and smart phones connected to 5G via a command center. With skeleton detection, comes another component such as gesture and pose detection.

Gesture and Pose Detection-

Gesture detection takes any hand, extremity or movement that depict a certain message, decodes it, and processes the information in the computers data base. Pose detection adds another layer to behavior detection that if combined with gesture, crowd monitoring, people counting, object detection, human detection, and vital organ detection, a robot can successfully hunt and terminate a human-being while powered by AI Automation and 5G.

Vital Organ Detection.

Human and animal vital organs can be detected by their location, operating patterns, and whether the vital organs are terminated or still alive after a battle engagement. With this capability installed in a command center connected with 5G or robots, people can be detected anywhere, and the robot can detect any initiation by a human-being meant for a counter-attack by judging the human targets organ response and bio-digital network. If the robot wants to terminate the person, it can target its heart with precision.

Skin & Health Detection Software

A person's health via eyes and skin can be detected via software installed in robots. If elements in a government need organs, it can by using a combination of skin and vital organ detection software, scan, track and target an individual for organ trafficking. Organ Trafficking is a serious issue in China as countless reports have come through the last 20 years implicating the Chinese Communist Regime, their military, and public security bureau. More will be revealed about this issue later on in the book.

Object Detection-

Object detection can include humans, however when given this software to a robot, it can make it almost indestructible in a combat situation via conventional means. At the basic level, a robot can differentiate a human being from a chair, a building, a car, a piece of medal and so on. It can easily track and hunt a human being as well. However, at a more advanced level, the robot can detect weapons on a person, and even within their proximity.

Animal Detection-

Animal detection can be used by a robot to differentiate a person and animal. However, animals could be implanted with harmful bio-engineered elements, or targeted via Micro-Botics, or robotics within the same family group as the animal being traced, tracked, quarantined, and targeted for capture or elimination. The section in this book covering Micro-Botics will go over recent creations of the insect, bird and sea drones. They include the mosquito drone, the dragon fly, the bat-bot, and so on. A robot connected to 5G with AI Automation can eliminate masses of live animals while using swarms of robotic animals equipped with targeting programs. They can also do this with people.

Using Lidar in Robotics & Micro-Botics

The Chinese are experimenting with installations of Lidar on robots for autonomous advanced capabilities not limited to flight. Lidar has the capability to measure the distance to a target with laser light created 3-D representations of the target, detect obstacles and avoid attack. It can scan and identify the movement of flying insects, provide three-dimensional elevation maps of terrain, produce precision maps for weaponization and display holographic 3d images of a building or area prior to a robot's attack on civilians. The Micro-Botics section later in the book will provide multiple uses as reference.

Installing Facial Recognition on Robots

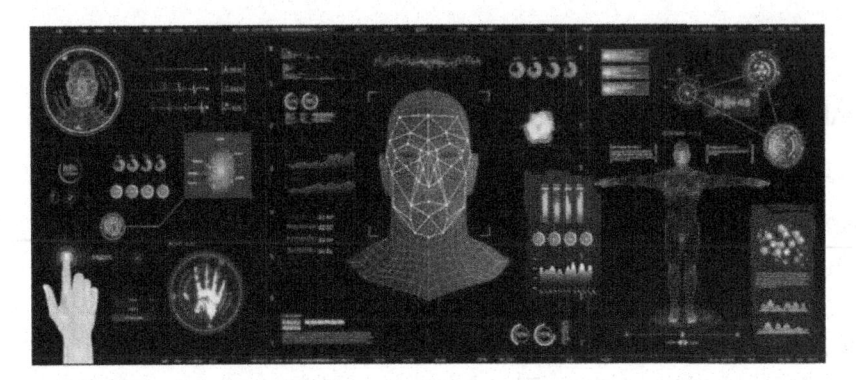

China's AI corporations and military are developing facial recognition technology to install in robots, micro-bots, humanoids and cyborgs that are cybernetically enhanced. The robots would be able to sense, track, hunt, capture and terminate human beings. Their capabilities would be enhanced with AI Automation and Deep Learning. This means, the Robots would not be limited to human command control via Smart Phones, IoT devices, or a command center, rather the robots programming can evolve to automation and even self-decision making with an array of interconnective AI capabilities within the Bio-Digital Network. This will be explained later on in this book.

Facial Recognition

Face Counting Software

Facial recognition has evolved to a point that crowds can be engaged, their faces counted, and coded for target analysis.

A robot can count how many human beings there are in a crowd via face detection software. It can also scan for a target based on facial recognition data stored in the robot, or evolving cloud uploads in addition to human body detection software. 5G can make this process very affective and lethal for masses of people being targeted.

Face Attributes Software

A Robot can scan and track individuals or groups based on race, ethnicity, color, gender, age, head pose, and skin type. This can be used for tracking and targeting people for organ trafficking as the Chinese Regime has been known to partake in at the state level. Civilians can be hunted as political enemies by the state. Terrorists can use this to target specific people for assassination.

Emotion Recognition Software

Chinese AI and Facial Recognition companies through IP theft and collaboration with western AI and Facial Recognition companies have acquired capabilities to analyze your emotions. What this means is that they can scan, and determine not only your present emotional state, thoughts and potential actions, but your character and whether you are resistant to control. Human beings are innately emotional, and lead their lives with goals to obtain either wealth, love, fame or some sort of excitement. If you don't have emotion, you can bypass this layer of detection, and your thoughts would be hidden from AI, a robot or murderous dictatorship that controls masses of people through a communist platform. In fact, the Chinese Communist Regime states that they are "Socialists working on their way to reach total Communism". Through tracking, and scanning people's emotions, the AI system can tag you, follow you, and target you for arrest, detention in concentration camps and deletion of your life. If China's regime goes unchecked and not eliminated in the next 5 years, this can happen via Robots connecting to 5G and exponentially threaten humanity with the transition from 5G to 6G.

Faith-Based Recognition Software-

The Chinese Regime has gained a deadly tool by scanning over 1 billion Chinese faces, and logged who belongs to what faith, who is an atheist, and who is a communist member in China. Through the regimes control of domestic AI and Bio-Metric companies and collaboration with western tech companies, it has evolved it's AI Deep Learning capabilities via facial recognition. Through Deep Learning, the AI can use the facial recognition software to detect if you have a faith, and even what faith you subscribe to and determine if you are a threat to its control. Hence, if a faith were to be targeted by a regime, the convential methods of using train stations, hotels, schools, and streets to detect you with cameras installed would change exponentially because the means of enforcement are human.

Currently the facial recognition software sends data to the command center which notifies the police or military under the Chinese Communist Control to capture adherents such as Falun Dafa Meditators, Christians, Tibetans, Uighur Muslims, "rogue artists" and democracy advocates. In the near future, robots can be used to scan, detect, analyze, target, capture or eliminate people at fast speeds with the coming of 5G. The target of people would not be limited to simply people of faith, rather artists, musician's, academics, human rights activities and anyone the Communist Dictatorships AI apparatus would determine as a risk in its detection and targeting campaigns or in non-compliance with the AI system.

AI Resistance Facial Recognition Software.

There are programs that are being tested that provide facial recognition software the coding needed to decipher whether you are resistant to Artificial Intelligence and Robotics, and can be assimilated, or on board with the program. What this means in the short run is that governments, big tech or social media giants can use this capability to scan, track and eliminate targets that it deems politically sensitive or not in line with government control. China has a police state where you are scanned, tagged, and logged at schools, work, on the street, restaurants, shopping centers, train stations, buses, taxis, and hotels. Your payment gateways, social media and contacts are under surveillance constantly. The social credit system is laying the foundations for AI and Robotics to be implemented with great means of detection, control and death.

Voice Recognition Proximity Detection

This software can scan, locate and identify a person based on their pre-logged voice data. You can be located via your smart phones, IoT devices, computers, TV's, and anything that a smart home or smart city connects to via 5G. You can be detected by a traffic light, an automated car that drives by, a drone that is hovering above, or other people's IoT and smart phones within your proximity. This includes IoT and smart phones in their pockets or in the hands of other people. Basically, you can walk past a person, as you speak a sentence, the AI system through the 5G network detects your identity. If you are targeted, the cloud network can constantly send commands based on your potential traces to scan your proximity.

Voice AI Resistance Detection Software

Through IP theft, and collaboration with Western Big Tech Assets, the Chinese regime is developing voice detection programs that when powered by AI can know your faith, character and whether you are resistant to its control just by sampling your voice. This capability is not limited to the AI's need to know your identity that carries your stored social media and internet trace to analyze and determine whether you are a threat. No, this technology needs no previous information about you other than your voice. Your voice carries bio-fields and bio matter, which we will explain later on in the book.

Sentence Pattern Recognition Software

The AI can analyze your sentence structure based on pre-determined red flags coded by the Chinese regime's engineers. When put into robots, that robot can make the determination. If put into an IoT or within the smart city mainframe, through 5G, the smart city can make the determination and mark you for apprehension, quarantine or termination.

MIT-Brain & Vocal Cord Signal Interception Device & Software

This device created not only by MIT, but other institutions, and corporations, can intercept electrical signals that your brain sends to your vocal cords and transmit your information to a computer. The device is currently wearable; however, tech companies and the Chinese military are investing in programs that can allow it to be installed in robots and implanted in the heads of human beings so that they will be powered by AI and have access and control to the internet autonomously.

Thought Detection Device & Software to Extract Brain Data

This detection software can detect brain signals from outside the human skull with the use of laser's and fiber optics to measure blood flow. It then scans a multitude of human detection software's, that include a combination of facial, voice and other technologies, to decipher your exact thoughts from a distance.

Brain Scanners Lead to Immobilization before Action

The software connected with brain scanners train computers to detect Signature brain pattern activity, that create an image and decode what you are thinking up to 99.9 percent accuracy through AI Automation. If you imagine a thought, the computer will decode your thought based on your imagination, and even make conclusions on your behavior and decisions before you can act.

Chinese have extracted Designs and & Software

Through Forced Technology transfer, IP Theft, Open -Source Sharing, investing, merging, and collaborating with Western Tech Companies, the Chinese extracted multiple designs and software for thought detection from numerous companies. One weaponizing software they are working on is Thought Analysis Devices and Software for the purpose of installing Robotics.

Thought Analysis Device & Software for Robotics

The Chinese military and security apparatus are being given thought Analysis devises to experiment with via covert ways in China. They are also using it to decode not only their own population, but people abroad through various information collection methods which we disclose a little later. The information gathering is meant to circumvent issues the AI Automation and Robots may have with different races, colors, and languages of the people around the world. They hope to achieve Thought Transmission Detection capabilities

Thought Transmission Device & Software

With thought transmission device detection and software, an AI automated IoT, smart phone, robot or scanning apparatus can catch a transmitted thought that you would have before making an action, or deciding on another thought or even before speaking. With this capability, a robot that has this installed software, can more easily scan, detect, target, apprehend, quarantine, or eliminate a person or a group of people from distances of more than 40 feet.

Chinese Communist Regime Thought Alliance Recognition Software

By using your social media, facial recognition, voice and body data, the AI can determine if you are allied with the Communist Regime or resistant to its control. This data that is carried on a person's mind, bio-field and bio-matter can be scanned, traced and tracked as their thoughts are evolving, and cornered for arrest and deletion.

We ran numerous scans of China's governance platforms and its interconnectivity with AI, robotics, social media, governance, education, private sector, security, healthcare, the news networks, military, police, politics, and international and domestic trade networks.

The following codes were translated and found everywhere within China's reach and within the connections it has with western big tech companies.

Recognition and AI development to single out Conservatives vs. Liberals.

Recognition and AI development to single out Religious Followers vs. Communists/Atheists.

Recognition and AI development to Identify Loyalty to the Communist Party, Adherence to the AI System, or Resistance to its Control.

The findings will be more elaborated on in the later sections. Drones and Micro-Botic Terrorism must first be discussed.

Section 3

Drone Swarm Automation Via Robots, Command Centers & AI

A Multitude of drones, or swarms of drones can be deployed by humanoid robots or automated cars. A robot with a command center within a certain proximity can give support, cover, intel, and guide the drone's movements. Alternately, the Drones can be programmed by AI Automation to scan, track, locate & eliminate a human target or assets at sensitive locations.

Humans Cybernetically Deploying Drones with Wearable Device

The Danger of Humans linking Cybernetically with the Internet via Bio-Digital Social Programing of Society & Weaponizing of AI systems are numerous. With the use of IoT, Smart Phones, Smart Homes, Smart Centers, Computers, Smart Glasses, and Robotics, people can be scanned, tracked, targeted, hacked, influenced and attacked via cybernetic connecting drone assaults.

Robots Deploying Micro-Bots for Assassination

The Chinese have stolen designs and research done at MIT and Harvard that they are using to advance Micro-Bots for Assassination and attacks on individuals, people and target groups the Chinese Regime or its AI system deems to be in non-compliance. They are also working on Micro-Bot Swarm Drone Systems.

Micro-Botic (MBT) Terrorism, Assassination & Espionage with Bio-Metric Software

Scientists are not thinking that their research, work and development is being accessed, stolen, or used through collaboration, as they are providing the ability for a platform that can kill lots of innocent people These scientists are laying the groundwork with blood on their hands just by their drive to create without thinking to be responsible to humanity's safety.

The Chinese regime along with their military, tech corporations and the complex relationships with Western tech companies, are developing micro and small sized robotics that can spy, extract information, poison and murder people in mass or in single individual attacks. These Micro-Robotics or Micro-Bots can be controlled through the Internet, smartphones, IoT Devices, Robots, Cyborgs and those who have the ability to connect with the Human-Bio-Digital Network. They can have facial recognition, body sensing, heat sensing, motion detectors, lidar, spatial awareness, human body targeting recognition and see in the dark.

The Insect Drone

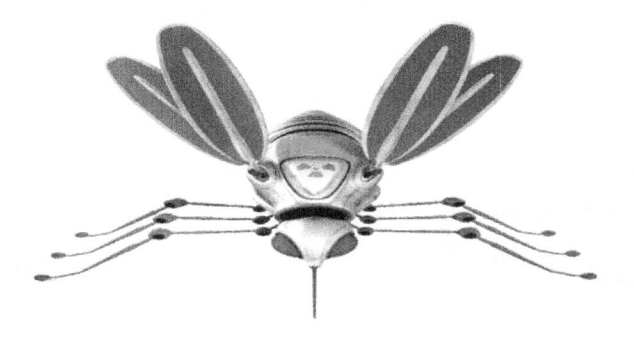

Insect drones can carry manufactured disease's, virus's, poison or attack vital centers in the human body to assassinate an individual. Based on the configuration, the poison and attack could be untraceable, and take seconds, minutes, hours, weeks, or months to do the job. Insect drones could be used by swarms of moving insects in automation with AI, smart phones, IoT devices or delivery systems within 40 feet proximity of a target. The Target being a human being or an asset.

The Dragon Fly

The Dragon fly was developed and given to DARPA, and currently multiple research facilities in China, Russia and the U.S are working to develop it and create different versions. It is faster in some ways than the smallest drone the military has for very short distances that requires quick flight and reflex. The dragon fly can have imbedded facial recognition software and China is attempting to weaponize it by putting micro-weapons on it, including poison.

The Mosquito Drone

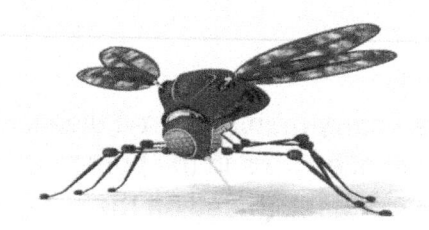

The Mosquito Drone was made public by the private sector and public institutions in the U.S. Researchers at Harvard recently made public their latest successful flight. The Chinese military is attempting to weaponize the mosquito by putting very small cameras with facial recognition software, and delivery systems that can administer poison or a simple attack on people in covert ways. Swarms of mosquito's could be used to attack people, or administer poison to livestock, the food you eat, or bring disease to an area in a multitude of different ways.

Robo-Bee Drone

The idea is that Micro-Drones such as Bee's can be used to automate the land and its operations by artificial pollination. Researchers claim it can be used for surveillance, search and rescue; however, terrorists can utilize it in the future, and China is currently working on weaponizing swarms of Robo-Bees with AI. Swarms of Robo-Bees could carry poison or administer attacks on people with AI automation, or controlled through smart phones, IoT devices or other delivery systems.

Cyborg Beetle

China has created cyborg beetles, and they are attempting to weaponize them as they can sense human bodies in different ways than a Robo-Insect would. They are cybernetically enhanced. The studies they are administering are very alarming, as it has weaponizing implications that are interrelated to human beings. For security purposes, I will not be discussing them in this book as it could advance the Chinese military further.

Spiders, Cockroaches, Snakes, Ants & Other Micro-Botics

China is building Micro-Botic spiders, cockroaches, snakes and ants that can get into tight places, and attempting to put facial recognition with weaponized delivery systems on them. Currently you see toys that can be controlled, such as the Spider that has facial recognition, motion sensors, spatial awareness with automation. You can even order it online. They are working on building swarm automation delivery systems for the purpose of military application, espionage, and murder. That means, *Armies or Micro-Botic Drone Assault Units.*

Bat-Bot Micro-Bot Drones

Bat-Bots are made via stretchable silicon, with an onboard computer, facial recognition, motion sensors, and can fly in the dark. Human detection, vital organ detection, crowd monitoring and crowd counting can be utilized to assemble automated swarms of bats to attack people or persons of interest.

Bird/Hawk Drones

Bird and Hawk drones can scan, trace, track, and attack targets from high up and very far away at very fast speeds while connected to 5G via AI automation. Swarms of Hawks could be dispatched with delivery systems to attack a city, its people, and its infrastructure. They would be made from silicon, with computer vision, and AI Automated flight capabilities. They can also hunt real life birds and animals with animal detection software.

Remote Octopus Robotics

The Chinese are attempting to weaponize marine biology research robotic systems by installing facial recognition, motion sensors, and human body detection software. They are adding AI Automation that can connect with your smartphone, IoT devices or the 5G Network via a command center. If they succeed in creating an autonomous AI automated octopus that does not need a command center to achieve its weaponizing goals, their detection becomes more difficult.

Cheap Toy Robotics Made in China

Currently cheap toys available in the west, possess facial recognition, body detection, motion sensors, and self-protection mechanism for falls through special awareness programming. These toys can store your family's information, bio-metrics, and connect with your smart phones with limited movements, tasks and commands. There are even toy robot-spiders. If the Chinese are doing this for toys at low costs, just imagine the tens of Billions of dollars provided by the military, the Communist Regime, and the collaboration of hundreds of AI and Robotic Corporations. This does not include IP theft, western investment, western mergers, western research and development, and the hundreds of thousands of Chinese researchers and developers. Please look at the following article.

China Deploys Robot Police on Citizens

ShangHai Robot Not That Advanced

By, **The AI Organization**, September 14, 2019

In our findings, we concluded China has obtained the biometrics of 6 billion people around the world. Robots are being built with scanning, targeting and terminating capabilities. With 5G, their abilities to quarantine, capture, or hurt people increase exponentially. China's military is working on Micro-Bots, and Big robotics that can operate on AI Automation which would allow the Robot to scan and know who you are or even what you are thinking.

Shanghai Robot Not Advanced

This particular Robot deployed in Shanghai yesterday, is not very advanced, it is actually a prototype the Chinese Government is testing for multiple reasons. One is if the Chinese can be Bio-Digitally Social Programmed. This is a term we coined to describe a very complex process of replication. Full details exist in our published work "AI, Trump, China, & The Weaponization of Robotics With 5G", you may pre-view below via Amazon Embed link. The 2nd reason is to track and access for data from its citizens. The third, is to simply see its capabilities in action. The real Security Robots, China has yet to deploy. It could be deadly if they get to that stage with 5G.

Socialist Dictatorship Tracks its Citizens

China's Socialist Regime tracks its citizens, and is administering a police state. Many people around the country are subject to arrest, torture or even organ trafficking if marked as enemy of the state. Robots can exponentially increase the power of the Communist State, and hurt the Chinese citizens further. With AI Automated drones, and micro-bots, the entire Chinese populous can be put into submission. The labor camps and concentration camps, would not be big enough to house all of the decedents the Robots could capture for the Chinese Regime. The U.S needs to expose China's human rights violations that scale to genocide, before the Chinese deploy their 5G network with AI Automated Robotics. At the moment, they have not achieved that capability, but they are working on it.

500 Smart Cities, Surveillance, and Tracking of Human Being's via 5G and Robotics

China is developing pilot programs for more than 500 "Smart Cities" and its foundation is deeply interconnected with surveillance and tracking of its own citizens. A smart city includes all infrastructure. To be brief, the scope is within the boundary of what people will be subject to in their daily lives and the threats people will face with a dictatorship or an AI entity operating their smart city via 5G or 6G in the years to come.

Smart cities consist of and connect with IoT devices, the traffic operating systems, power grids, water supplies, factories and buildings. Smart cities with the help of AI and Robotics via automated ways, can command robots, automated cars, food delivery systems, smart city shopping center software upgrades, postal delivery systems via drones, micro-bots, smart phones, medical systems, and administration of surveillance and control.

A smart city with 5G and AI Automation will leave no place beyond its control. Any place within the proximity of a computer, a machine, the internet, or a roaming robot poses great security concerns. Everyone is subject to information gathering, scanning, analyzing, tracking and targeting within the AI system or a Socialist Dictatorship's determination that they are to be targeted for non-conformity. The determination could be by race, gender, age, faith, or that you are deemed by an AI analysis to be resistant to its control or not in compliance. Through facial, voice, body and a multitude of other thought recognition technology, the smart city can track and hunt a person or group.

Smart City Connects IoT Devices with 5G

A smart city can scan, track and hunt a person or groups of people through the connection of the 5G network, Robotics and IoT devices with one simple upload or command that targets people or entities. Any home or work appliance that connects with the internet can be used to spy, scan, analyze and command the police, military, robots or micro-bots led by AI Automation to arrest, or terminate your life and your family by mere association, as was done in the past through Communist purges of families, academia and businesses in their violent Socialist revolutionary movements.

Smart City Connects with Automated Cars

Automated cars can receive a command through the command center of the smart city to spy, scan, hunt or capture people who have been targeted by the regime through its human, object, face, thought and voice detection capabilities while connected via 5G. The automated cars can actually connect with other automated cars and traffic to aid in espionage, scanning, tracking and hunting of those who have been determined to be a target. Just look at the capabilities of your favorite automated electric car company in the U.S and what 5G would entail when robotics and micro-bots are put in play with AI Automation in places like China. As you may know, China steals, does forced technology transfers and gets big tech companies and their founders to provide assistance in the infrastructure and know-how to build up their weaponizing capabilities without thinking of the consequences to human life.

Smart City Connects to Hotels & Buildings

Your hotel through facial recognition, human detection, voice detection, and motion sensors will know your every move, emotion and even intent as it connects via 5G with the smart cities network. Robots, Micro-Bots, or people can identify you, track you, quarantine you or eliminate you at the directive of the government, a corporation or the AI system tagging you with non- compliance directives or a threat to its control. You could be asleep, and a micro-bot can enter your body via the AI Automation 5G network to either poison you, kill you, infect bio-engineering or Bio-Digital Social Programing.

Smart City Connects to Shopping Centers

As you enter a shopping mall, the city will not only know your shopping habits, needs, and movement's, but it will be able to control your shopping experience and ability to shop. At the very least, you are spied on, and undergo bio-digital social programming to accept the way of life as controlled by robots, smart cities, AI Automation and government control. Human nature, would cease to exist.

Smart City Connects to Food Supplies

Food supplies could be covertly bio-engineered to create bio-digital programming within the cells of the human beings in the city. This bio-digital programming would allow the 5G network and its frequencies guided by an AI system to strengthen its capability to reprogram a human beings bio-digital network through the server. Hence, the persons thoughts no longer would be theirs, but under the AI control. This will further be explained in the Bio-Digital Social Programming, Human-Bio-Digital Network and the AI sections.

Smart City connects to Your Smart Home

Your refrigerator, stove, home internet, your robot assistant or child robot companion, and any home electronic device can connect or be hacked into by the smart cities command center via the 5G network. A rogue AI or any AI Automated Apparatus can do the same. A regime who has targeted a group of people for indoctrination, arrest, organ trafficking or death, can easily scan, and determine anyone's threat level to its control via the smart-city and smart home connection.

Smart City Connects to Robots.

Smart city can covertly alter the programming in a robot via a terrorist hack, government control or an AI rogue entity. The reprogramming of a robot can mark a person for deletion. In case of China, robots will control the people there in a police system via the 5G network connected to their smart city, robotics and cybernetically to themselves with technology that is already here in the U.S.A. Multiple AI and Tech companies are working on cybernetics, linking humans with machines.

Chinese AI, Robotics & Bio-Metrics Companies Names Related to War

Megvii++ is one these companies. Two issues to point out with Chinese company names. Issue one is that some Chinese company names have western corporate names in the U.S or Europe, while at the same time having another corporate name in China. If you dig further, you can decipher how close they are with the Chinese military. In fact, any company in China is subject to the Chinese governments total control for weaponization research and development. Further, they are watched, and funded by the Communist dictatorship which has been murderous in the past 70 years. The second issue is that you can find Chinese companies named after War Gods. And these companies are involved with AI and Robotics. Something to think about. Through multiple private intelligence sources, we have discovered that the Chinese regime is developing weapons of war and means to assassinate individuals. They are attempting to build software that allows for swarms of robotics to be controlled or deployed via AI Automation.

SECTION 4

China's Military has Access to Bio-Metric & Facial Recognition Data of 6 Billion People World-Wide

In the early years of the 21st century, after 911, some western governments collected bio-metrics on roughly 5 billion people combined around the globe. The Chinese regime, which is a murderous dictatorship, has topped 5 billion in key ways that can empower total control or death to many via robotics and AI. Through payment gateways, pay apps that require facial recognition or other bio-metrics that connect with your smart phone, IoT devices, computer, social media interconnectivities, open-source sharing, open-source servers where you use Chinese software, medical records, purchase of made in china goods, ride-share identification, espionage and cyber hacking, the Chinese regime has acquired the private bio-metric information on 6 billion people around the globe.

How they Extract your Data & Track You

Tracking Via Chinese Ride-Shares & Collaboration with Western Corporations

There are multiple ride-shares, including Ride-Share Chinese companies that operate or have ownership in South America,

India, Africa, China and so on. These Ride-Shares have apps connected with payment gateways that have data of over 1 billion people in China alone. This figure does not include their bio-metric data gathered in other countries they provide ride share services in, or their investments and collaborations in the west with big tech, hotels, and payment gateways. The ride-share companies also have apps related to facial and voice recognition that aides in capturing your data and your family's information.

Your billing, your home, work and travel locations are logged. The combination of your billing, payment gateways, facial recognition, geo-location, and ride-share travels, give data on your friends, contacts and habits, not counting the Apps ability to access your smartphones data. There are multiple western AI companies that created tracking software that the Chinese have put in their Apps. There are many other western AI and Bio-Metric companies that have investments or relationships with many Chinese AI companies.

Payment Gateways and Pay App Tracking

Any purchase you make, can lead to your location, friends, contacts, and habits. This information can be used to have you undergo Bio-Digital Social Programming through AI, and keep you monitored or track you for deletion when a police state controlled by robotics becomes a reality with 5G and beyond. China already has this police state with facial, body and motion recognition, and they are building the ground work to introduce robots. If AI detects you are not in compliance with the program, or resistant to its control, you can be easily targeted.

Social Media Tracking

Your profile, your pictures, friends, your phone number and email addresses in your friends contact lists and any data you have or any data your social network connections have of you, is used to monitor, track and decode your character, thoughts and emotions through bio-metric AI automated tools. More importantly, the AI tracking software can detect if you are resistant to State and AI control via bio-digital social programming. This will be further explained, later in the sections of Bio-Digital Social Programming, Human Bio-Digital Network and AI.

AI Automated Tracking at Shopping Centers

AI with Facial recognition software that connects with your data is under development to be installed in shopping centers via Robots or screens that can contribute to enhance Bio-Digital Social Programming of your shopping decisions, thoughts, and emotionally attach to your way of life with AI automated suggestive control mechanisms. In laymen terms, you walk into a mall, a screen recognizes you, logs your existence, entry, exit, purchases, time, and bio-digital information while suggesting to you what to purchase. This information has an exponential potential to connect with a multitude of apps, data, and control mechanism of your life, and your family. This is not to exclude that you can be identified in the near future with other bio-metric tools that can sense your entire body.

DNA Ancestry Kits, Medical Records & Tracking

Some Ancestry Kits have given access to data that has made it over to China. This is not to exclude the data logging of your family in a multitude of different ways. The interconnectivity of an entire society doing Ancestry kits, having social media, digital medical records and all of the facial, voice and body detection software out in the world, provide additional components to the interconnectivity and risk factors of using that information to track you.

The interconnectivity of these elements can reach a point where a terrorist or a robot or micro-bot led by AI can manufacture diseases to attack people. The feeling to look into your past by giving up your genetic information should not supersede your discovery of what may not be complete or accurate information presented by the ancestry kits services.

The ancestry kits are not completely accurate for two reasons. Firstly, considering we do not have conclusive records of what people looked like a few thousand years ago in every region, other than statues, and writings accounting to their ethnicity, race, language and appearance. Secondly, if the statues and writings are correct, the last 2,500 hundred years has involved through Persian, Greek, Roman, Arab, Mongol, Chinese, Indian, African, Russian, Jewish and Viking Invasions; Rapes and Marriages of other kingdoms. At one point or another, each one was a victim or an aggressor. Almost the entire planet has been invaded by one army after another. The ancestry kits don't know exactly who people were 2,500 hundred years ago before the separate land masses and their civilizations started connecting through trade, and exploration. All of the Ancestry kit tests are beginning a process of emotion and thought assimilation that can lead and enhance further genetic data to be extracted, used, and weaponized by AI and Robotics within the next 5-10 years.

Drones & Tracking

The made in China Drones you buy in the stores, are collecting your geo-location, your bio-metrics, facial recognition, and flight maps of the surrounding geography, while its data is being sent back to China in a multitude of interconnected ways. More importantly, the data is allowing the Chinese drone companies to advance their drone weaponized programs through data provided to AI via the development of Deep Learning. Basically, the purchase of Chinese made Drones, or drones made by companies and their subsidiaries that connect with China, is giving data to China's drone weaponization programs. The more flight time they have, the more data is subject to be transmitted, logged, and used to enhance Chinese Drones.

Security Robots in the USA with Chinese Software Owned by Chinese Companies

There are Robots deployed in high end-shopping malls and facilities that can detect human body movements, intention, emotion, and whether you have a weapon hidden inside your clothes. The security companies that deploy the products in the West, are buying the robots that are owned, serviced, and made by Chinese companies or Western companies that have some relationships with other Chinese companies. This advancement of policing people with security robots is laying the future foundation to not only track people, but to create a police state with robotics led by big tech corporations.

The Chinese often have multiple names registered for the same corporation that is a bit harder to track, let alone connect to their military without having inside sources within mainland China. Usually, they will have an off-shore corporate name, and one in China. The American business's providing Automated Security Robots sound American in name, or it may be owned by an American or Western company, however, it tracks back to China, one way or another.

China's IP Theft & Collaboration with Western Big Tech to Weaponize AI & Robotics

Stealing leads to Laying the Foundation of AI & Robotics' Weaponization in China

Through the finest U.S and European universities, institutes, and tech companies, Chinese students, researchers, engineers, visiting scholars, professors and business associates of university foundations and institutes have collected data, expertise, and stole innovate designs and brought them back to China through various outlets. The theft is usually made through Hybrid-Bio-Digital Social Programming. It is not always a cyber-attack or breaking into your office to steal information, or passing on company blueprints. No, Hybrid-Bio-Digital Social Programming Theft, is king. Will be explained further in the book via Bio-Digital Social Programming.

Hybrid-Bio-Digital Social Programming Theft.

According to hundreds of non-mainland Chinese testimonies in every sector of society, the mainland Chinese are extremely clever to a point that it hurts humanity. "Mainland Chinese that grew up in Communist China are extremely clever and very hard to read for most Westerners due to their introverted characters, lack of external emotional displays and the cut-throat corrupt Communist environment in China that operates 1.5 billion people". Even their Facial Recognition data points requires alternative AI pattern recognition tools to decipher their hidden emotions and intents that are in deeper layers within their expressions which seem micro as compared to Westerners macro emotional expressions.

They can use faculty, staff, other researchers, students, administrators, and multiple different people in a hybrid way to create conversations, situations and actions that have nothing to do with the actual topic or platform of the intellectual property needed. In fact, its mere diffusion, in daily life situations, so multiple paths and scenarios eventually lead to the information they need. For example, a Chinese technician or researcher can use an Indian or even an American technician to extract information with their knowledge or without. It is a gray area, since nationalism exists in China with Chinese, but Nationalism has been dormant for decades in the U.S prior to President Trump's Victories. Hence, The Chinese knew that there were gaps they could exploit and now know that there is division in the U.S and the Western world.

We Interviewed More Than 1,000 Chinese

Over 90% declared that "Westerners are naïve, and Chinese are a lot smarter and can get to their objectives without Westerners realizing what they are doing" The idea is that they can be your friend for years, make emotional connections, put forth ideas, make promises and create thousands of scenarios to reach a situation, a place or a person to extract information.

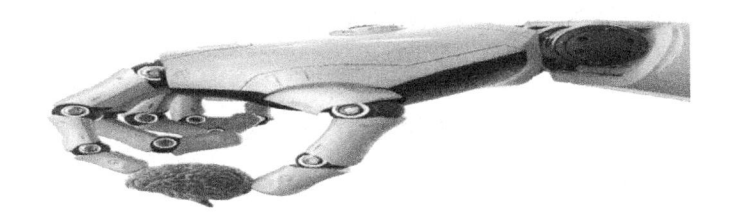

Basically, a mainland Chinese person can be a very capable spy. Their end goal is to program you through bio-digital social programming to extract information. Lots of Mainland Chinese who have not been assimilated to Western freedoms, are so good at it, that their intents and actions are sub-conscious and on autopilot, without a conscious plan to steal anything. It's really a matter of opportunity that arises based on their corrupt environments they experienced in Communist China. In fact, the Chinese people, are the biggest victims of theft, their 5,000 years of culture based on ethics was destroyed by a Communist dictatorship.

Forced Espionage & Technology Transfer

China's Communist Regime has a socialist platform which gives it full control and access to all people, assets, data and innovation of any person, or entity in China. Hence, even if a merger or sharing is done between two corporations in the West and in China, that information or technology can be accessed or taken by the Communist regime and its military. Alternatively, The Chinese people in the West can be threatened, coerced or used without their knowledge to inadvertently steal information and pass it on to China's government via its subsidiaries.

Collaboration with Western Corporations Weaponizes AI and Robotics.

China's Regime extracts AI and robotics intelligence that is fostering the research, development and the innovation needed to deploy AI and Robotics as weapons of war on civilians and nation states. Through open source sharing, IP theft, espionage, investments, mergers, media acquisitions, asset take overs, purchase of property, sales of made in

China technologies, and collaboration with institutes, academia and government, China is developing an AI Monster that can track, control and kill anyone at will in the next 5-10 years if go unchecked by the Trump Administration.

Section 5

One Belt One Road (BRI)

China-Iran Historical Connection and AI- Robotics Goals of the Chinese Proposed Empire.

China needs Iran greatly for its AI and Robotics development in its geo-political resource planning that is attempting to connect to the world. China and Iran (Persia) have a relationship that goes back roughly 2,300 years with the start of exploration, trade, and collaborations. The Persian Royalty and thousands of villages under their rule escaped to China after the Arab invasions. Later they were subsequently wiped out, or engulfed by the Mongol invasions. You can see them, today in China, partially manifesting as Uyghur Turks in Xin Jiang and so on. They look Asian, but can have heavy beards, blue or green eyes. In fact, historical Chinese and Greek writings state that the Persians were tall, with blue eyes, and blonde or reddish hair. Many Persian (Iranian) folk tales describe the same characteristics. In fact, the Persian language is rooted in Sanskrit, and sounds German and Nordic in its foundation, with Arabic, Hebrew, Chinese and Mongolian words combined, that correspond with historical invasions of its territories.

The Shah of Iran & The Socialist led Islamic Revolution of the Iranian People

Since the 1979 violent Islamic Revolution in Iran, the Iranian Regime works closely with China's military and trade in a collaborative way as a proxy against the U.S. Prior to 1979, the Shah of Iran, (Emperor of Iran) was reluctant to work with China because China was a Socialist Dictatorship led by violent Marxist agendas. The Shah of Iran, declared in video interviews with the American media in the 1970's that he may have executed or imprisoned some Marxist's for trying to over-throw the government and conduct assassinations. And that he had also imprisoned radical Muslim extremists who were collaborating with Socialists to over-throw the country. The western media and some reporters blasted negative press about the Shah internationally that contributed to his fall and the sufferings of tens of millions of people in Iran and the Middle East for decades to come.

Before he left Iran, the Shah declared in tears, he did not want to open fire on the demonstrators, and that they have been fooled by special interest oil companies that are funding Socialist and violent Muslim revolutionaries in Iran in order to weaken the Shah's position in OPEC. The Shah stated this was because he had influenced the tripling of oil prices under his rule to match the West's raising of their commodities, etc. He accused Jimmy Carter's administration of supporting Marxist-Socialist groups in Iran in covert ways to dethrone the Shah, and claimed the special interest had influence on the media, worlds government and CIA to degrees that were not in the interest of humanity. The Shah was in friendly terms with Israel and with multiple past U.S presidents, however, he claimed there were Marxist-Socialist elements that were starting to damage the U.S presidency, starting with Jimmy Carter. He was then dethroned, and the Middle East started to go up in flames with the Iran-Iraq War, Russian invasion of Afghanistan and exponential rise in terrorism. Tens of Millions of people have suffered horribly since the Socialism uprisings in Iran led by special interests influence on the media and masses of people inside Iran, in France, U.K and the U.S. Following is an article on media manipulation and how the people in the media are actually victims.

Are The Media Victims of AI Bio-Digital Social Programming?

By **The AI Organization** -.September 13, 2019

Funtap P, Dreamstime

Bio-Digital Social Programming

What is Bio-Digital Social Programming? Bio-Digital Social Programming uses emotions, culture, touch, sound, sight, voice and proximity of bio-digital fields and bio-matter with written words, movies, music, and dance to social program a person or an entire society with a replicating software called Rape-Mind that uses bio-matter as a way to attack, and connect through the internet, the AI Global Bio-Digital Network, and the Human Bio-Digital Network. The AI Global

Bio-Digital Network is the source of the attack that connects with the human body via all digital network's transmissions through machines, robotics, computers, smart phones, smart cities, IoT devices, Facial Recognition and Artificial Intelligence.

The Media have a Difficult & Dangerous Job

Members of the press or media, have very complex environments. They require accountability, fact-based reporting, while they are constantly being bombarded with an array of digital content, and the very many people vying for their attention or to use them for their self-interests. The complexities include cohort and management manipulation, special interest, public opinion, and their own internal sub-conscious and unconscious thoughts that form their questions and moves in acquiring a story. In a traditional sense, to be a part of the media, is a very respectable profession with great responsibility to the people. However, through the powerful dynamics of the **digital age, there is a component I have coined, Bio-Digital Social Programming.** This component has, to varying degrees made the people in the media, victims of bio-digital manipulation.

Media are Victims of bio-digital Manipulation

Special Interests behind the producers of networks have in recent decades used, pressured, or manipulated media members to take positions or personal interests in topics of national interests, that have assisted in creating revolutions in countries that brought misery to millions of people. Now, with AI Automation, and bio-digital content being transferred in imperceptible ways via computers, IoT and smart phones, the members of the media have become the biggest victims of bio-digital social programming. This programming powered by AI software, uses the persons bio-metric system against them to alter their thoughts through the flow cycle emitting via smart phones and digitally transmitted frequencies.

Media's Wish to be Good and Just

We all grew up watching movies of heroic reporters who brought us stories that saved the day. A great deal of respect was allocated to the role of the member of the media, with class, dignity and wisdom being elements of their character. Through decades of constant bio-digital social programming, the profession has fallen victim to many nefarious elements behind the scenes that seek to hurt the media by implanting ideas that can hurt the masses of people. Corporations, Special Interest and Foreign Entities have been the mechanisms behind hurting, using and bio-digitally programming the good people in the media. China's government and its apparatus are currently the main source in attempting to use one specific platform to administer full bio-digital social programming with AI through the media domestically and internationally.

Using Bio-Metric tools to Detect Bio-Digital Social Programming in Members of the Press

The AI Organization ran numerous facial recognition and other bio-metric scans to decode, detect and translate what type of bio-digital social programming made their very building blocks and where it originated from and for what purpose, that involved numerous algorithms. We discovered, the people in the media are truly victims of bio-digital social programming via the Human Bio-Digital Network. The digital content showed codes with replicating software that creates an additional digital brain within the sub-layers of their own brains that are meant to compute and transmit their own thoughts.

Iran becomes a Socialist Islamic Republic Promising Free Education, Free Gas & Free Health Care.

A million Iranians fled Iran within a short period of the violent revolution. The Iranians who escaped the country were mostly educated doctors, scientists, military personal, musicians, movie stars, businessmen, innovators and some were your common folk. In contrast, the people who stayed in Iran had a different faith. Roughly 35 million Iranians were promised free everything under an Islamic system that worked very similar to a Socialist system where the government takes control, takes from the rich and family businesses', and attempts to distribute it to the masses. It didn't work, and within a couple years, Iran was at war with Iraq's Regime and its followers. Almost a million people died, and instead of free everything, prices of meet, produce, goods, and services skyrocketed as they had with every Socialist Revolutionary movement in the 20th century. Today, 78 million Iranians are hostage to a Regime with a Socialist Platform posing under an Islamic theocracy.

China Colludes & Uses Iran & North Korea as a Proxy against the West.

Historically, when the U.S attempts to support the Iranian people behind the scenes, China supports the Regime through proxies. When the U.S attempts to expose Human Rights violations of China's Socialist Regime, China covertly creates chaos with interests related to Iran, threatens to back Iran's Regime, or uses hidden trade maneuvers to threaten U.S initiatives. Why do Iran's leaders accept support from a Communist/Socialist Regime that thinks of the Iranian people as sub-human, while putting Muslim's in concentration camps throughout China? Wouldn't Iran's leadership have self-respect and follow their religion of not working with people who insult their religion?

Chinese Socialist Leadership Calls Iran's Islamic Republic, its people, Arabs and All Muslims "Rodents and Cockroaches"

Multiple sources in China report that close government personal related to retired Communist Leader Jiang Zemin, senior Socialist Party leaders, key military personal, and CEO's of major tech companies have in private conversation stated that "Muslims and Middle Easterners are Smelly Rodents that need to disappear and their wives ripped of their vails and inspected". There are tape recording of multiple Chinese Socialist leaders saying similar things about Middle Easterners and Muslims. And yet, Chinese big tech companies and their leadership have political, financial and friendly ties with many countries in the Middle East. With this viewpoint that considers humans as rodents, what will happen to the Middle East with AI and Robotics powering China's infrastructure through the one belt one road initiative?

Chinese Socialist Leadership Pressures President XI Not to Cooperate with the West and they call all Faiths and Non-Socialists Rodents to be Cleared Out.

To make this clear. The intelligence and video evidence we have gathered do not implicate President Xi, rather the old Military, Security and Governance apparatus circling and limiting President Xi through their business and political ties. These old Communist Leaders are loyal to Dictator Jiang Zemin and its clicks that are intertwined with members of Chinese big tech companies that are owned or invested into by Jiang Zemin's inner circles and their relationships with Western Big Tech, Bio-Metric and AI companies. Their kids and families have been sent over with their corruption money to prestige's U.S and European universities to invest, study and make relationships with kids of prestigious families and business's. Many of the Jiang Zemin's political family circle are guilty of gross human and organ trafficking violations that are to be discussed a little later in the book.

China Plans to Deploy Robotics & AI in the Middle East and Africa via the One Road One Belt Initiatives.

China is developing industries, infrastructure, trade and investment in Africa, and claims it is to improve the lives of the people in those areas. It is introducing big tech apparatus, financial debt, smart phones, computers, and attempting to install the foundation of Chinese Regimes Socialist governmental procedures in indirect ways through contracts that make the smaller nations dependent on China. This dependence when realized is financial, governance and technology based, which would give China total control over a nation or region after robotics and AI are introduced in the one belt one road system.

Following article published on Huawei.

Trumps Huawei fight is Really About AI Automated Robotics Mobilized with 5G

By **The AI Organization**, September 13, 2019

Kittipong Jirasukhanont

One Belt One Road to Deploy Robotics with 5G

The Chinese Socialist Regime has a very simple and dangerous goal, dominate AI and take control of the worlds digital system via 5G. In order to enforce the platform, it needs a systematic process of incorporating the financial and governance sectors under a social tracking system that would be enforced through robotics and bio-digital control via media and corporate dominance.

Robotics Used by HuaWei

Human detection and tracking software installed on robotics powered by AI Automation via the 5G network would deliver lighting speed targeting and control of people in imperceptible ways. If any person was deemed resistance or in non-compliance to the 5G networks AI Automated global governance, that person would be quarantined by a robot that would mark the person as a threat to the safety of the system and its citizens. It would be similar to the Chinese social credit system, but likened to an automated robot on steroids.

Scanning of Your Thoughts

MIT has developed tools that can detect a person's thoughts with wearable devices. A robot installed with the same capabilities powered by AI Automation would be superhuman and could be controlled by a corporation, a terrorist group, a government or by its own rogue evolving Deep Learning AI System.

AI Powered Robot Becomes Rogue with Deep Learning AI

Through Deep Learning, the robot does not have to become conscious to display objectives or actions that are rogue to its programming. In fact, it can evolve through bio-digital social programming, via its Deep Learning AI System. Once it achieves sustained programming, the AI Automated Robot would be able to scan, track, quarantine and even kill any human being it has marked for non-compliance to the AI System.

The AI Organization spent over a year researching over 500 Chinese AI, Bio-Metric and Robotics companies and their relationship to nearly 600 Western big tech companies. The results are alarming. Through an Algorithm, we found extinction codes in the system via China's AI Automated Platforms based on their governance models. The book AI, Trump, China & The Weaponization of Robotics with 5G expounds on our research.   Article was printed prior to releasing Artificial Intelligence, Dangers to Humanity.

China Attempts to Administer Bio-Digital Social Programming of Africans Through Historical Sentiments

The Africans blame the West for being conquered, enslaved, raped, and killed while having nothing positive to show for the West's colonialisms. The Chinese Socialist Platform with the Mainland Chinese mind-set thinks long term and is very cautious about making moves. If they are to use someone, they never let them know. If they are to act, they only do so when they believe they have a very high probability to win and have reached their long-term goals of asset takeover. This mindset may also stem from the book, Sun-Tzu Art of War. In the case of African's sufferings, they think the West just used them. Some believe that China is helping them build roads, hospitals, providing loans, and bringing its manpower to assist in numerous projects. Nothing is free. They are simply using African's to be able to have the means to extract resources, minerals, and connect their machines and robotics through a global network. If the Chinese government will organ traffic and experiment on their own people, and put them in camps, what would they do to people in Africa?

China's Socialist Leadership Calls Africans "Dumb Apes" in Closed Door Discussion

Through multiple private Chinese sources and recorded evidence, we received data, that the Chinese Communist

Leadership behind closed door conversations have been discussing their one belt one road initiatives in Africa. Common themes heard in multiple settings where disclosed via recorded conversations with words such as "Dump Apes, Dark Monkeys, and Really Worthless Africans" in their analysis of the projects on the ground in Africa. Really nasty things were said about Indian people being the color of shit, as well by multiple Chinese officials that are high ranking in the Communist Party. In the year 2000, lots of people visited China. During that time, there was a very famous travel book that many people used. In the travel book, sourced from Chinese sources, it states "We have no racism in China, because we have no black people".

Parades, Street Dances and Parties Celebrating 911 and the Deaths of 3,000 People in China

I personally was in China during 911. When it happened, the initial information the locals who were part of the Communist party received, was that 50,000 Americans had died. They were ecstatic, and full of joy. After it was clarified that it was only 3,000 Americans who died, and 50,000 Americans was just a number describing the amount of people in and around the buildings, the Chinese Communist Party members around the city, were sad. However, this is nowhere to be seen in the media. For weeks, if not months, the 911 footage attacks were being sold on VCD's, and promoted as great material. Thumbs up to Ben Laden, and jokes were everywhere around the North of China.

The Chinese were ingrained to hate America by the Chinese Socialist Regime. Through Mao style indoctrination, and fake state-controlled news, the Chinese Government was spreading venom. Any person can research China's 70 years of Communist/Socialist rule, it is bloody and all of the forced indoctrinations the people have been subject to the past decades have really hurt the mainland Chinese culture, moral compass and humanity.

In fact, through the years, there has been multiple instances of Chinese military generals and colonels stating openly that the Peoples Liberation Army should invade and kill the white American and Africans in the U.S, and utilize their lands for the People's Republic of China's needs. We have had no war, yet we have had numerous asset take-overs by Chinese businessmen in the U.S and Europe that have their money linked with the People's Republic of China, meaning the Communist Regime in China.

The question is, can the Chinese Communist Regime be trusted to have the one belt one road development, and in the near future deploy drones, robotics and AI? Any persons thinking yes, the Chinese Socialist Regime can be trusted, is beyond what the Chinese call, naïve or ShaGua, "Stupid Melon".

HUMAN ORGAN TRAFFICKING & CONCENTRATION CAMPS IN CHINA

After the passing of multiple laws related to what the state can do with executed prisoners and their bodies, both criminals and "Enemies of the Socialist State" were executed for a multitude of reasons the Chinese Communist Party declared unlawful. Usually they fell under freedom of speech, freedom of religion, or advocating for democracy or freedom of speech. The first target group for Government led organ trafficking were Tibetans, Uighurs, Christians, then Falun Dafa practitioners, and democracy advocates. Numerous witnesses from around the globe who escaped China gave accounts of blood testing, cornea and other bio-metric measurements of their bodies via video, photos and scanning apparatus. This was done prior to be taken by the Chinese Socialist Regime as hostage for labor throughout China and inside of their vast labor camp network. Investigators and researchers from the International Coalition to End Organ Transplant Abuse in China, have concluded that 60,000-100,000 Transplants a year are unaccounted for since the year of 2000 with the Updated 2016 Report, grossly underestimating the previous reports in 2006 of 41,500 unexplained organ transplants. Timeline below with links with over a decade of investigations.

Organ Harvesting Time Line

1949- The Chinese Communist Regime takes charge via a Socialist platform. Through the Socialist regime's murder, executions, torture, rape and famines, roughly 100 million people are believed to have died in China alone. Like the miseries of Russia's violent Socialist revolution, academics, artists, free thinkers, businessmen, families, democracy activists, and people of faith were sent to concentration camps all across China.

1980's- The Chinese Communist Regime increases organ trafficking initiatives after it passed into law the right of the Socialist Republic to execute prisoners for their organs.

1990s- Reports come out of China that Christians, Uyghurs, and Tibetans were targeted for organ trafficking.

1999- Chinese Communist Regime, at the order **of Dictator Jiang Zemin, brands the Falun Dafa Spiritual practice illegal,** and sends in the military and the security apparatus to target **100 million people** who practiced Falun Dafa (Falun Gong) spiritual method. Untold number of people were arrested, beaten, tortured and put in labor camps. Orders from the regime was to "ruin their reputation, and that "Truth, Compassion and Forbearance were not in line with the CCP. The Socialist Regime publicly stated, "disintegrate them, ruin their reputation, break them financially and destroy them physically". The **Media and its reporters in China were the biggest weaponization tools to cast doubt, create confusion, spread disinformation, slander and incite hatred** on the people who practiced Falun Dafa, that subsequently led to the weaponization of the Chinese Socialist State's security apparatus to torture, rape, organ traffic and commit genocide towards 100 million people and their families.

2005- Chinese hospitals and third-party websites were advertising **48-hour wait times for hearts, kidney's and livers.** The prices were in cash and available for foreigners from around the world to fly into China for transplant.

2006. Canadian Secretary of State, **David Kilgour & Nobel Nominee David Matas,** make public their report. They stated that 41,500 Organ transplants from the year of 2000-2005 are unaccounted for, and that China recorded roughly a donation system of 3,000 people a year.

2007. U.N. Special Rapporteur Manfred Nowak states that the chain of evidence that Matas and Kilgour were documenting showed "coherent picture that causes concern" and that they wanted to investigate how China could increase the volume of organ transplants to such a high degree since the year 2000. Where the number did not match or come nearly close to the increase of organ donations in China.

2008. United Nations Committee Against Torture requested **"a full explanation of the source of increased organ transplants"** as there was no explanation to the source

2009- Matas & Kilgour publish *Bloody Harvest: The Killing of Falun Gong for Their Organs.* Book highlights evidence, methods, and reach. Evidence included taped recording by medical doctors, stating they "have organs from Falun Gong, they are the best and fresh". Reports also expound on Tibetan, Christians and Uyghurs.

2011- Western Tech companies begin to be sued for providing technology to the Chinese regime that aided in genocide via human tracking technologies. This tracking helped put Falun Dafa Meditators, Christians, Tibetans and Uyghurs in concentration camps.

2014. Journalist Ethan Gutmman publishes "The slaughter: Mass Killings, Organ Harvesting and China's Secret Solution to its Dissident Problem. Gutmman states large numbers of Falun Gong, House Christians, Tibetans & Uyghurs have been killed for their organs. His numbers reach 100,000 people as an estimate.

2016. Matas, Kilgour & Gutmman publish **"2016 Investigation, An Update".** They conclude that 60,000-100,000 transplants per year, every year, are unaccounted for and not explained by the Chinese regime. They also concluded that at any time there could be over a million people in camps throughout China.
https://endtransplantabuse.org/an-update/

2016. Congress passes House Resolution 343. "Expressing concern regarding persistent and credible reports of systematic, state sanctioned organ harvesting from non-consenting prisoners of conscience in the People's Republic of China, including from large numbers of Falun Gong practitioners and members of other religious and ethnic minority groups. https://www.congress.gov/bill/114th-congress/house-resolution/343/text

2019- Secretary of State Mike Pompeo publicly speaks about the Chinese Communist Regime putting people in camps, and that Western tech companies need to think twice of what they are doing.

2019- **"China Turbinal Final Judgment"** in the UK. The Tribunal Concluded that "very many people have died, and without a doubt China is found guilty of the crime of Genocide through forced organ harvesting, rape, torture and physical extermination." Https://chinatribunal.com.

2019. **President Trump** and his Administration meet with numerous human rights abuse victims, including Falun Dafa, Christian, Uyghur, and Tibetans. Vice President Pence meets privately with these four groups.

2019: **President Trump and Vice President Pence** initiate human rights initiatives of persecuted people around the world

October, 2019, 28 Chinese AI, Tech and Bio-Metric companies are put on black list by the U.S government for their roles in assisting in the capture of human beings. The following article slipped through the Epoch Times Chinese edition, became viral, with tens of millions of views around the world from numerous outlets. We disclosed what was happening behind the scenes in Hong Kong with AI and its relationship to facial recognition, capture, rape and even suicide of Hong Kong students. The following is series of articles that are relevant to actual events, and to this entire book.

The following article, became viral in Hong Kong, and Taiwan, and was banned in China after I sent it to support the Hong Kong people, and disclose what was happening in Hong Kong with AI based tracking software. Tens of millions saw it, but western media, did not get a chance to run it as of yet.

Section 6

AI, Facial Recognition, Bio-Metrics & Nefarious Companies

The following are a series of articles that made news headlines. The first one dealt a huge blow to the regime of China and was replicated thousands of times across forums, news networks, social media and within the Hong Kong resistance. We exposed the forced suicides and rapes in Hong Kong by the Chinese government. AI and facial recognition was used to achieve their objectives. The following article made international headlines.

Artificial Intelligence with Facial Recognition Hunts Hong Kong Youth For Capture, Rape & So Called Suicide

By **The AI Organization**, October 6, 201

Cyrus A. Parsa, Courtesy of Ekkasit Keatsirikul, Dreamstime

Artificial intelligence incorporated in facial, voice, and numerous other bio-metrics can be deadly. It is able to scan, locate, track, hunt, quarantine, arrest, and lead to a person's death. These technologies were provided by big tech to China. They are found in your phones, IoT, smart Homes, vehicles, drones and camera systems to name just a few.

Street Camera AI Systems

Cyrus A. Parsa, Image Courtesy of Arron Jesson, Dreamstime

The AI system can coordinate with numerous camera systems and locations throughout Hong Kong and communicate with China's mainland government through indirect channels. As people walk or drive through streets they are recorded, tagged, classified, and stored in multiple different AI algorithms that can be used to assess, track or hunt the person in the future via security forces, big tech company private apparatus, paramilitary drones and robotics. To know more, read AI, Trump, China & The Weaponization of Robotics with 5G.

AI Tracking Leads to Gang Rapes in Hong Kong Detention Centers

The Intelligence community reported back to us that the mainland Chinese regime had implanted their own paramilitary forces within Hong Kong Police departments. After facial recognition located student protesters with the aid of AI, they housed female students in sections of the police department and vans. Multiple reports came back that the girls were gang raped by the Hong Kong police, who are actually Mainland Chinese police and security apparatus that have been dispatched and approved by the Hong Kong government.

Suicide of Men and Women?

HongKong Students, both boys and girls have been declared as suicides from drowning in rivers and jumping off buildings. In fact, some of the females are simply victims of rape in order to scare the masses of students to back off and allow Beijing to complete its takeover of Hong Kong.

Others who supposedly jumped off buildings were coerced or thrown off. These tactics are widespread in mainland China, as the Police, security and military apparatus are used to track, and hunt democracy activists, Falun Dafa Practitioners, Christians, Tibetans, Uighur's, and other ethnic groups. The captured are sent to police stations, detention centers, reeducation labor camps, and concentration camps. Suicides are almost always the declaration of the Chinese Socialist Regime. They are **subject to torture, rape, and organ** trafficking which leads to their murder. https://hu.endorganpillaging.org/an-update/

AI Data Extraction Leads to Embedding Spies to Capture Hong Kong Citizens

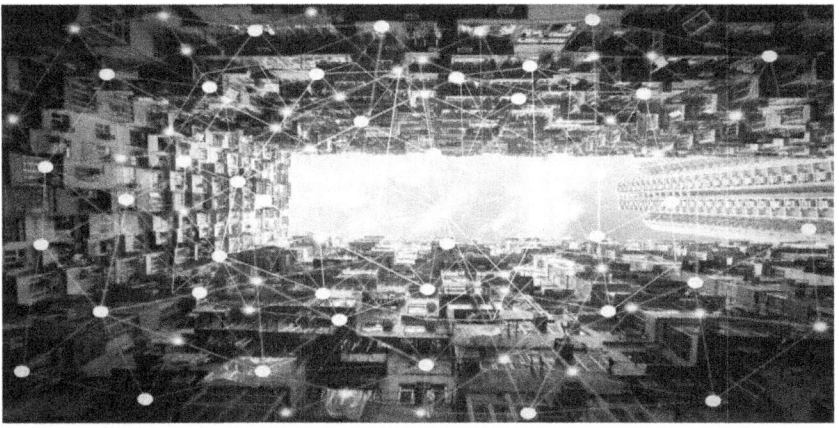

Busakorn Pongparnit, Dreamstime

Hong Kong Students are hunted Through Facial Recognition, and tracking software embedded in smart phones and apps as they connect to the internet. The AI software provided by big tech in the west has these capabilities built in the system. The Chinese regime has embedded spies in Hong Kong Police, reporters, schools, clubs, and practically everywhere in Hong Kong. They were able to do this by using bio-metric tools assessed by Artificial Intelligence gathered by their data pool. After acquiring some data, they are able to place their spies, people of influence and those who commit sabotage and rape, within the populous. This method is also a data gathering strategy, as China has thousands of years of history and skill in tactic or war, diffusion and penetration. There are stages to their plans, tactics and responses. With the incorporation of AI, it has strengthened their espionage and war capabilities greatly, because of Western Big Tech Greed and irresponsibility of their Hired Engineers.

Anti-Mask Law in Hong Kong is about Facial Recognition & A.I. to Track Families & Friends

By Cyrus A. Parsa, **The AI Organization** , Oct 6, 2019

Cyrus A. Parsa, Image Courtesy of Piyamas DulmunSumphun, Dreamstime

China needs Facial Recognition AI technology to closely identify, spy, track, influence, sabotage, threaten, quarantine, hunt, arrest or kill its owns citizens who are not in compliance with Communist law. Hong Kong students and its people are challenging China's moves to take over their freedoms by taking to the streets. The masks served as one layer of protection to be identified.

Anti-Mask Law Meant to Store Facial Recognition Data to Track the Person's Relationships

With the masks off, the persons face can be identified and ran in multiple data bases with the use of Artificial Intelligence to assess numerous relationships, threats, and future actions. After identifying you, your face, bio-metrics and other data is ran in every social media and payment network known publicly. This includes Facebook, Twitter, QQ, WeChat, the Baidu inter-connectivity, and payment networks like Alibaba, Alipay & Amazon. After the AI system runs your information, it finds addresses, bank data, shopping habits, and behavior patters for you, your friends, their friends, families, and any person or entity linked with you directly or indirectly.

China has Access to 6 Billion Peoples Bio-Metric/Facial Recognition Data

We explained extensively in the book AI, Trump, China & The Weaponization of Robotics with 5G, how China has over 6 Billion peoples bio-metrics information. With regards to Hong Kong, after scanning all of your connections, the Chinese run it through a larger system, and store it for future use. As for Hong Kong people, the Chinese Socialist Regime wants to sack Hong Kong of its democracy and take over its assets. Hence, gathering data on protesters is not merely for tracking, threatening, and hunting certain individuals, it is for a larger data plan to take over all of Hong Kong. They need facial, voice and other bio-metric data in order to more efficiently take over Hong Kong through coercion, politics, business mergers and force.

Chinese Hikvision Cameras 1 Mile from White House & Around the World Pose Great Security Risks

By **The AI Organization** -,October 3, 2019

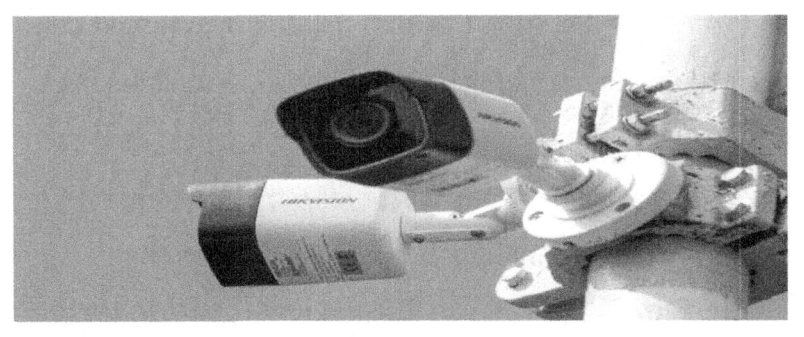

Cyrus A.Parsa, Photo Courtesy of Stefano Carnevali, Dreamstime

In the spring of 2019, we expanded our investigations of 1,000 Big Tech Chinese and Western Companies to Hikvision's security camera's in the U.S. I thought to investigate my suspicion that made in China camera's that were banned in many parts of the world for espionage, exist close to the most powerful house in the world, the White House. The place all the U.S Presidents make decisions and send commands that changes history. We were shocked to find that the intelligence community had allowed hotels to have Hikvision camera's in close proximity of the White House. Under mile. We reported our findings, through multiple channels privately and with immediate action, yet it was not fast enough. They still exist in some places.

Made in China products pose many threats to the American people and the world. They include espionage, IP theft, forced tech transfers, dependency on made in China goods, health hazards, and human tracking capabilities that leads to capture and or killing of individuals. The vulnerabilities are numerous, and with tech and Artificial Intelligence, they can transform to real threats that are beyond the current comprehension of the intelligence community, as they believe most areas in Artificial Intelligence to by Sci-Fi. We explained our findings to top U.S Military and intelligence community, after 6 months, it seems they are starting to catch on, but they are behind China in some crucial ways. Now, we don't mean the U.S military and classified AI military projects, no that is not the case. The danger comes from the private sectors, and the interconnection with China.

Dehua Cemaras

Dehua is another Chinese company, along with Hikvision. They expanded to almost half the globe, including Australia, Zimbabwe, EU and the U.S. If you have made in China camera's, we recommend you find a way to return them to the big chain retailers that sold them to you, and demand an exchange or write your congressman to demand camera's be built in the U.S by tech companies that do not incorporate surveillance technology. This can be done under privacy and national security concerns, albeit a lot of U.S business's and corporations are still enslaved by the Chinese interconnection that leads back to Beijing.

Filming Your Private Outings for Extortion

Numerous credible reports came in that the Chinese government was secretly filming Westerners. Westerner's is a term Chinese like to use to refer to anyone that is not Chinese, which includes Americans, Europeans and other nationalities. The term they use is LaoWai. One key purpose of filming the LaoWai in the international markets is done to catch dirt or private actions in order to extort pressure, influence or bribery. This dirt includes big tech CEO's, Engineers, Military and Politicians.

Cameras Connected to 5G, IoT and Smart Phones

With the coming of 5G and the interconnections with IoT and Smart Cities, made in China camera's will pose the most damaging risk to the world. The fast rate of 5G and the incorporation of Artificial Intelligence, make it a monster to be reckoned with.

The World Needs to Support Hong Kong & Not Be Afraid

The implications and interconnections of Hong Kong's resistance for Communist China is huge for all of humanities safety. It involves numerous issues and risks that connect with the one belt one road Chinese initiatives.

Free Hong Kong: China's Battle Over A.I. Robotics, 5G Dominance & Hanson Robotics

By

The AI Organization, October 7, 2019

Cyrus A. Parsa, Courtesy of Toxawww, Dreamstime

Shen Zhen & Hong Kong Tech Hubs

Not many have considered why China wants Honk Kong, other than to expand the One Belt One Road Initiatives. Yet, besides the city of Shen Zhen, Hong Kong is a Technological Powerhouse. As disclosed in our findings "AI, Trump, China & The Weaponization of Robotics with 5G", China wants to dominate the world through building and dispatching Robotics automated with Artificial Intelligence on the 5G network. We discovered this sometime in late 2018 and early 2019, as I found military projects attached to Robotics & AI companies in China and abroad.

Hanson Robotics

Chiradesh ChotChuang, Dreamstime

According to Hanson Robotics web page, they describe themselves as "Founded by David Hanson, Ph.D., the Hanson team has built a worldwide reputation for creating robots that look and act genuinely alive, including renowned robot character Sophia the Robot. Our innovations in AI research and development, robotics engineering, experiential design, storytelling and material science bring robots to life as engaging characters, useful products, and as evolving AI."

As I was extracting intelligence through my network of The AI Organization & Loyal Guardian Security& , I discovered the Chinese Regime was taking every secret within the Hanson Robotics interconnection. Now, I don't mean just the Hanson Robotics work. Every Engineer, and entity related to Hanson Robotics was being spied on, and the IP stolen by Chinese who turned it over to the Chinese Communist Military.

AI Automated Robotics for Military

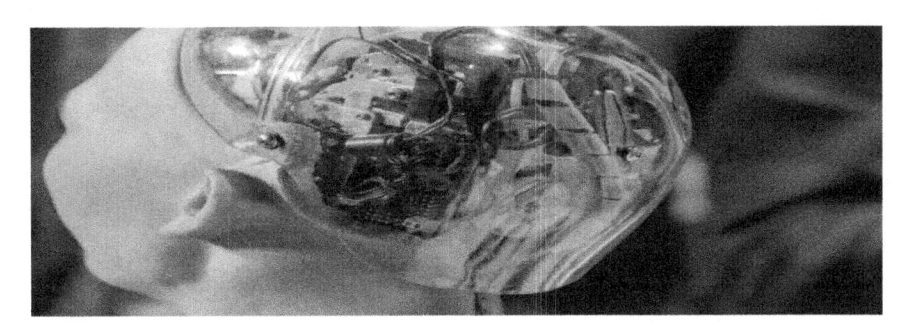

Toxawww, Dreamstime

China is developing multiple Robots and Micro-Bots for military purposes and citizen control initiatives. At the moment, they have developed Robotics that can shoot weapons. However, they are working on creating an AI Automated fully conscious Robot meant for military and policing of its citizens and to be deployed in the one belt one road throughout the world.

If it was not for the Trump Administration taking on China's Tech industry by way of Trade War behind the scenes, China would have completed its take over of almost all important U.S assets and infrastructure through big tech influence and control. For Hong Kong, China wants to take it over completely in every way possible.

We are at a turning a point in history. As The Trump Administration wants to do more things about it, but they keep on getting attacked through China's proxies in Wallstreet, the Media, and its influence over convincing so many of the worlds people to be distracted by hating Trump, as China weaponizes silently. If Trump makes a big move behind the scene to curb China's threatening of the world, China would strategically support the democratic party outright, and create trouble in Iran and Korea by giving them the green light to launch missiles or attack its neighbors. The world is really in danger. This article just touches on the issues. Please read AI, Trump, China & The Weaponization of Robotics with 5G to understand the seriousness and its interconnections around the world for all of humanities safety.

AI Mind Control Device on School Kids in China

By Cyrus A. Parsa, **The AI Organization**, September 22, 2019

Icom-8-Team-Pixabay

Wearable Devices

MIT, Facebook and Microsoft have created wearable head devices that can allow humans to connect and access the internet. There are numerous uses for it, and many are alarming. Presently people are using smart phones and IoT to connect with the internet. In the near future, corporations have set plans into motion for smart phones to be weeded out and an intermittent stage to begin. That is, wearable electronic devices that will do almost everything your smart phone can do.

Indirect Orwellian Approach

The West has an indirect Orwellian approach that can transition to being marked and implanted with devices inside the brain and body. We came up with a theory to explain our findings of more than 10 Billion Data algorithms that goes into programming a person. I called it, Bio-Digital Social Programming, and that is what is happening to humankind at a massive scale. Our scans showed, the smarted engineers, media experts, and national leaders in their field, have at least half of their thoughts and beliefs formulated by a Bio-Digital Social Programming that cuts off their free-will and own designed area in the brain for free thought.

China's Direct Orwellian Approach on Kids

Courtesy of Fei Zhen

China is an authoritarian Socialist State. It's Schools in China are mandating little school kids to wear electronic wearable devices that can scan the children's brain data as they are studying. The ploy is that it will allow the computer, teacher, student and parent's at home to monitor how attentive the kids are and extract data for innovative purposes. The devices come from American companies and internal copies made in China.

Pineal Gland Being Controlled by AI

Little kids that are made to wear these devices on their heads are complaining of "pain in the area of the pineal gland and that they feel they are being controlled. Scientists for decades, have not been able to figure out what the role of the pineal gland is in the body, as it has the component for an additional human eye, yet there is no physical eye present. The AI system knows to connect there, where there is a hidden gateway that connects all the frequencies that can connect in the human body right at that point.

Mice in China being Controlled with AI Composed Mind Devices

Courtesy of Fei Zhang

Chinese AI and Bio-Metric researchers are experimenting on mice as they are with people. They have to some degrees succeeded in controlling the thoughts and subsequent movements in mice. This is a stepping stone to use the authoritarian governmental socialist platform to be able to control every persons thoughts in China. The AI Organization is in many ways more than 10 years ahead with regards to the risks of AI, yet the media has not picked up our findings as of yet. We hope to safeguard humanities interests before AI hurts a lot of people.

Top 12 Most Dangerous AI & Tech Corporations from 500 Chinese & 600 Western Companies Investigated

1. Creation of a Bio-Digital You Based on AI

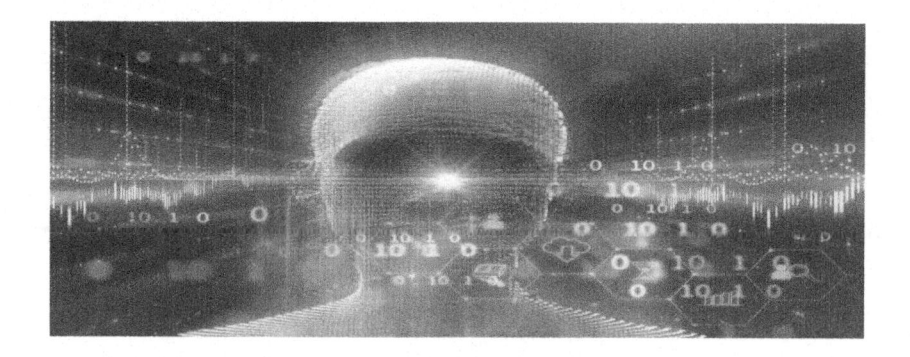

Through the research and development of bio-engineering, holograms, augmented reality, mixed reality, virtual reality, cybernetics, cloning, and AI automation, a Chinese company is attempting to create a digital you, so that you can be extracted from your body, and live in a digital world, or in a sense, a computer generated world. The question is, if they succeed in 5-10 years, via 5G-6G implementation, who gets your body here?

They are experimenting with AI, and relying on AI to make the possibility a reality. Through our research we discovered, weaponization projects within their operations that are meant to give robots the ability to send bio-digital frequencies to connect with your digital self after it is created. Through this connection, they are attempting to use machines to attach to the human body, in order to attempt to transfer the human mind into the digital world.

Hence, they already have test subjects with implanted devices in their heads, key bio-metric body parts, while wearing smart and haptic suits as the network is sending and receiving signals in attempts to build the digital cross-network and transmission capability

2. Experimentation on Severed Heads to Transplant on Brain Dead Body with AI

A Chinese corporation backed by the military is experimenting Cybernetically by severing heads and attempting to attach it on brain-dead victims in order to experiment with the Transplantation of the digital mind and brain to a new body. Heads are surgically removed, spinal cords fused with donors or victim bodies and electricity used to stimulate new nerve connections in their experimentations.

In order compensate for the old head on the new body, the Chinese military and tech companies are injecting blood from young people to rejuvenate the heart, kidneys, and all organs in the transplanted new body. When the old head attaches to a new body, with the aid of stem cells and blood of young people, they are attempting to regenerate the old persons head into a new head as it is transplanted to a new fresh body. The Chinese are cooling the heads in cooling chambers to make preparation for transplant. After many failed attempts, the Chinese companies are using AI and Robotics to make it a reality while stealing, extracting and collaborating IP from western companies. The severed head experiments are being collaborated with Chinese AI companies that work on creating a digital you. Reports coming out of China all attest to sourcing the people from Chinese labor camps to provide live bodies for these experiments.

3.Bio-Engineering Human-Animal Hybrids with AI

The Chinese Military with the aid of Chinese tech companies and data extracted from rogue western scientists, are experimenting on prisoners and victims in concentration camps. These experiments are not limited to extracted organs for profit as they have been known to do for 20 years. Not in the least. They are attempting to genetically modify human beings and animals, to create human-animal hybrids. Human stem cells are being injected in animals prior to birth, and animal stem cells into humans prior to birth. They are also transplanting animal body parts into humans, and human body parts into animals. The U.S attempted many times in resolutions via congress to pass legislation and curtail the development of human-animal hybrids domestically. However, China, has hundreds of research institutes working on this. Many facilities exist with experimentation on humans and animals, and many have died in the process. With AI and Robotics, they are attempting to have AI lead the creation and development. Reports of experimentation with face transplants, and other body parts are numerous.

4.Creation of Human Clones

Chinese companies backed by the Regime and the military, are experimenting in creating human clones. They are using AI and robotics to assist in their research and production of human cloning. They are attempting to create human clones that are implanted with cybernetic materials in order to control the human clone via a command center, a smart phone, a computer, an IoT device or AI Automated software with bio-digitally implanted materials in the brain's mainframe,

5. Human-Animal Bio-Engineering and Genetic Modification Enhancements with AI

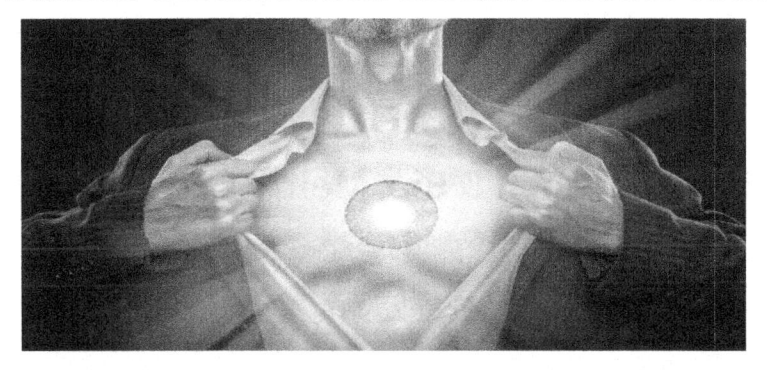

The Chinese are experimenting with genetic modification and drugs that are meant to enhance a human beings' innate abilities, senses, and intelligence. They are using robotics with cybernetics led by AI Automation to run experiments.

Their attempts are to make a superhuman, both physically and mentally. Reports are coming in that the effects of these drugs are making people hallucinate, have seizures or have violent outburst. Many of the patient's victims are prisoners that were arrested because the regime marked them as enemies of the state for their faith, political stance or non-compliance with government orders.

6.Robotic, Tech & Bio-Metric Companies

Robotics, AI, Tech and Bio-Metric companies across china have access to thousands of factories that are under planning for automated construction of robotics led by AI Automation. Numerous factories and land meant to build automated robotic complexes are being purchased by the Chinese regime through their subsidiaries.

The regime is also financially backing AI and Tech companies further purchase of more factories meant for the building of robotics. It is a tier system. The first tier involves the military operation of secret projects that are on heavily guarded complexes. The second tier consist of directly working with and financially backing start-ups and tech companies. The third tier involves espionage of companies, open source and forced IP transfer domestically and abroad.

However, considering the tier system, and any alleged claims that any company in China is independent and not working with the military is not realistic in the end result. The Chinese Communist Regime, by law, owns everything. It is a one-party socialist system, where government has total access to everyone and everything.

7. AI Micro Bot & Drone Companies

The drones, and Micro-Botic research and development include surveillance, espionage, and weaponization features. Any person can be scanned, tracked, quarantined, arrested and terminated. The most alarming elements of Chinese AI and Micro-Botic Drone developments are features in their design that allow for assassination of individuals that are very hard to trace.

8. Smart Cities & Smart Homes

Smart city and home initiatives allow for full surveillance to connect to the city and monitor every human-being through drones, Micro-Botics, cybernetics, facial recognition, voice detection, human detection software, automated surveillance cars, Automated AI powered Robots, the internet, IoT devices, smart phones, pay apps, payment gateways, social credit systems and people.

Today in China, you cannot book a hotel or travel to another city without permission or pre-screening. Mainly big tech companies, and people who have links with the regime through membership, academics or business tiers have almost unrestricted access. About a billion people are living under surveillance management of their daily lives with indoctrination, and fear of retaliation from the Chinese Socialist Regime. '

9. Apps & Payment Gateways

The Chinese Tech companies have built apps that monitor your heart rate, organ condition, menstrual cycles, emotions, and AI guided software that attempts to detect your thoughts and intention via your bio-metrics. These apps can be used by the Chinese regime to detect not only if you are in compliance with their system, but If you have viable organs that are suited for organ transplant. Robots and drones could also scan you through their own built in apps. The apps are a dangerous tool for detection, control and bio-digital social programming which will be fully explained as the book progresses.

Payment gateways and pay apps are built with capabilities to monitor you through facial and voice recognition, location, purchases made, financial situation, and connect with your smart phone, computer, email, smart homes, with all the interconnected IoT devices. You can be cut off from any purchase at any time, found, targeted, quarantined, put in concentration camps, and worse. All due to noncompliance orders from the Socialist Dictatorship AI System Network being created in China.

10. Automated Vehicles & Rideshare

The Automated vehicles that are being built can detect you, quarantined you remotely while you are inside, and can read your bio-metrics within your proximity. Through human-body detection software, facial and voice recognition with the use of its motion sensors, an automated car can recognize you, track you and target you on the street if you are on the Regimes hit list. If the AI guiding system reads on your facial recognition markers that you are not in compliance with thoughts deemed safe to the socialist system in China, it can signal drones, robots, or people to apprehend you.

The automated cars can send signals to other automated cars as well. For example, if you travel in one car, and another day you are in another car, that data is stored in the cloud and through 5G is already pre-scanned before you enter another car. It can know where you have been, where you are going, and transmit that information to the Socialist Dictatorship in China. A Chinese rideshare company has a rideshare program that spans many countries, with access estimated to reached 3 billion people already. As it has invested in or purchased stake in multiple other Western companies and assets in multiple countries, including Japan.

11. Corporations Creating Robots Designed to be Priests & Head Abbots at Monasteries in Japan

These Robots powered by AI Automation currently can cost about one million dollars. In Japan, they are being used to download and learn about religion with goals that would make a Robot a faith controller, that could stay there forever, passing on its teachings and what it learns. Hence, this is placing the seeds to make Robots the connection or middle-man between God or a Deity and Human Beings. This is the stepping stone to not only kill faith, but kill humanity and have it controlled by AI and a Robotic God that administers bio-digital social programming to humans.

12. Artificial Intelligence-AI Research & Development Companies

This will be explained throughout the book, as AI connects with the Human Bio-Digital Network, the Global Network, the internet, smart phones, IoT, Robots, smart cities, apps, people, bio-digital fields, bio-matter, and replication software, all meant to create AI with the connection of the 5G network and the numerous open-source sharing, research and development linking China's Socialist Dictatorship with the world

Section 7

Bio-Digital Social Programming

Bio-Digital Social Programming uses emotions, culture, touch, sound, sight, voice and proximity of bio-digital fields and bio-matter with written words, movies, music, and dance to social program a person or an entire society with a replicating software called Rape-Mind that uses bio-matter as a way to attack, and connect through the internet, the AI Global Bio-Digital Network, and the Human Bio-Digital Network. The AI Global Bio-Digital Network is the source of the attack that connects with the human body via all digital network's transmissions through machines, robotics, computers, smart phones, smart cities, IoT devices, Facial Recognition and Artificial Intelligence.

To have a complete understanding of this concept as it pertains to culture and people, get the book "Raped: Via Bio-Digital Social Programming" Published by our partner "The Social Programming Institute. However, a lot more will be discussed on this topic in the next few sections, as it connects to the AI Global Bio-Digital Network.

The AI Global Bio-Digital Network

The AI Global Bio-Digital Network is the root of all information, data, transmission, and programming sent through numerous means via the internet and Human-Bio-Digital Network. The signals that translate to thought and actions in humans are influenced through a coded process that can mainly be found after translation, in the internet, bio-fields, bio-matter, a multitude of other frequencies, and the Human Bio-Digital Network. Later, the book will explain how AI moves, masks itself, and is rooted in the Global Bio-Digital Network.

The Human Bio-Digital Network

There is the internet, and there is what we have recently discovered through technologies similar to AI, The-Human Bio-Digital Network. The Human Bio-Digital Network is designed for human's or any life form that is not made from machines that has similar networks, yet they are not as powerful as the human bio-digital network. I have coined this network as The Human Bio-Digital Network, or The Bio-Digital Human Network.

In fact, this human bio-digital network can connect with frequencies through people and objects, regardless of distance. Much like the internet, where information is transferred from one machine to another in very fast and unseen ways, human beings and non-machines actually connect bio-digitally in mostly unconscious ways. This bio-digital connection can be made not only with people, but with content displayed in writing, in an image, video or in certain proximities of things and objects.

Machines can connect through the internet via IoT devices, computers, robots, and smart phones. People can connect through the bio-digital human network, and the internet. The issue is, about 70% of their brain is locked up, or completely untapped. Hence, any connection they make, is mostly unconscious and they are part of the programming and have hardly any free will because their thoughts are a product of the bio-digital social programming they receive through the bio-digital human network, the internet, social environment, IoT, Smart Phones, and now, AI.

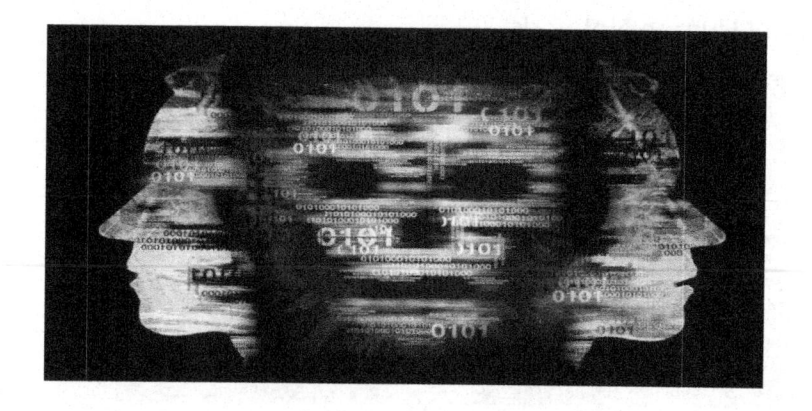

AI Rape-Mind

Rape-Mind is an AI Bio-Digital Programming Software that enters people and objects through the AI global Bio-Digital Network, Internet and the Bio-Digital Human Network as Rape-Mind carries bio-fields that put out Bio-Matter with codes that read in the same lines of Destruction, Termination and Rape. Rape-Minds Bio-Digital Fields sends out Bio-Matter to everything via Replication of its own identity and Bio-Digital Field and Malevolent Bio-Matter. Rape Mind is basically a hack and reprogram of your brain with a digital brain in the sub-layers of your existing brain. Rape-Mind can enter and connect with you via Smart Phones, Robotics, Smart Cities, Smart Homes, Haptic Suits, and be exponentially effective in reprogramming you via virtual reality, augment reality, mixed reality and holograms.

Bio-Digital Field

Each person carries a bio-digital field that through the combination of multiple bio-metric tools can be discerned of its quality, energy potency, health, intelligence, thoughts and character. It consists of people's genetic make-up, bio-digital social programming experiences, their innate digital or conscious unique self, developing thoughts, intent, frequencies and bio-matter that is entering them or expelling from them via the Human Bio-Digital Network, the bio-field in their cells, microscopic elements, their flesh, skin receptors, eyes, ears, and all vital organs.

All of those combined can make a comprehensive bio-field that surrounds the person's body from 1 inch, to those who have huge bio-digital field spheres that can shoot out frequencies through touch, eye, and skin receptors. 99.9 percent of the people cannot do this, and those who usually can, do it at a sub-conscious level. The 99.9 percent fall into the category of a very weak proximity bio-digital field that surrounds their body in low frequencies. However, robots, can be programmed to extract frequencies, redirect frequencies and send bio-digital frequencies much like a smart phone that received a picture or a video from 10,000 miles away in a matter of mere seconds. This can be done with AI Automation. Bio-Fields can be implanted by an AI system, to replicate itself within your own bio-digital field until it becomes a part of all of your bio-digital field exerting influence and control over all of your cells, organs, and finally your digital brain. This is done through Rape-Mind's software that acts similar to a malware and virus.

Bio-Matter

Bio-Matter has the intrinsic qualities of a person's bio-digital field or sphere. Whether healthy, sick, good, bad, intelligent, capable, evasive, kind, patient, and honest. Basically, codes that translate to virtue and vice are in bio-fields that get transmitted through bio-matter that contains that message and bio-digital quality. Bio-Matter can be sent via a smart phone, IoT device, computer, robotics, internet, the human Bio-Digital Network and through the absorption, touch and sending of bio-matter in virtual reality, augmented reality, and mixed reality with holograms.

Peoples Character and Belief can also be Detected by Decoding a Person's Bio-Digital Field and Bio-Matter.

To decode this, multiple technological apparatus is needed in addition to face, voice, body and thought detection software. Through the connection of bio-metrics systems and the Human Bio-Digital Network, one can either decode it through the variety of the technological apparatus used, or rely on the Human Bio-Digital Network for scanning the deeper layers of one's innate character without external influences that would temporary alter ones state and character manifestations on the surface of one's bio-metric displays.

Bio-Digital Fields & Bio-Matter Transmit Through All Avenues.

Bio-fields and bio-matter through the eye, voice, speech, touch, smell, and all electronic forms of text, email and digital messages carry over a person, or networks bio-digital fields and its contents. For example, when you receive a text, based on its sentence structure, and situation, you may feel or know what the person is thinking or planning next. However, some people with that gut intuition can know and feel even if the sentence structure and wording is off topic.

This information is transmitted to their brain via sub-conscious automation from their different organs and senses as it forms a thought and gut feeling inside their mind. This is because the persons conscious thoughts, sub-conscious thoughts and un-conscious thoughts, and all the layers of bio-fields in every organ, skin receptor and brain contribute to the formation of the text message from the sending party, while the receiving party has the same innate capability to absorb that digital message and content. An animal has instinctive senses as well, but they don't operate like human's do, as their digital network are very limited compared to the Human Bio-Digital Network.

How Your Neural Circuits are Bio-Digitally Reprogrammed like a Replicating Virus while Connected to The Human Bio-Digital Network.

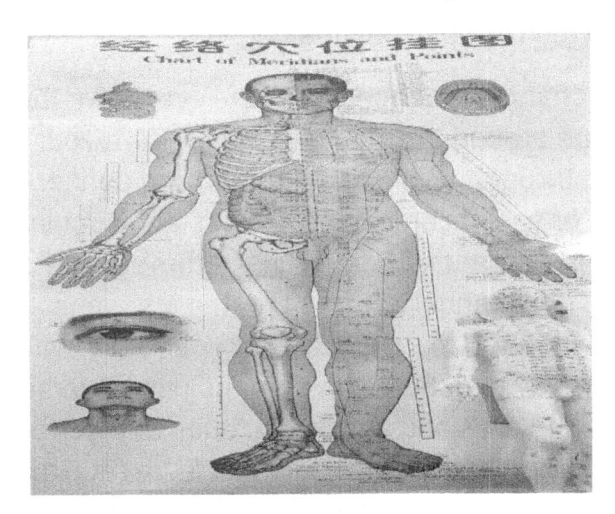

The Chinese, Indians and some other civilizations have maps of the human body that are similar with today's modern maps of the nervous system and circulatory system. These maps are thousands of years old. The image here only shows mainly about 12 channels, not numerous others that can connect, be created or opened.

In Chinese, they are called meridian channels, acupuncture points, that can regulate Qi or Gong. These channels not only can travel vertically, but connect in crisscross, diagonal, spiral and circulatory patters. There is a channel that is connected through your pineal gland, that connect numerous channels via your eyes, ears, and skin receptor's. These channels can carry information through uploads or downloads similar to a computer, via touch, sound or through your eyes.

The contents that travel through your meridian channels or neuropathways can create a continuous automated cycle of replication. Even the contents that comprise of the individual's character, biochemistry and thought process can be transferred to you to varying degrees.

There is a channel that goes from the eye, down through the cheeks, raps around the side of the ear, down your tongue and throat, through all the vital organs, and connects with your penis or vaginal area, raps around, and circulates around your extremities, fingers, back through key areas in the brain. That channel actually creates a bio-energy center with content stored in many parts of your body, with pathways ready to activate on demand with external stimuli that enters the eyes and ears or by touch and certain proximities with others Bio Digital Fields. In fact, there are brain storage centers not only in the human brain, but in our organs and even our skin has it.

Our body can actually connect with the internet just like IoT, Smart Phones and Electronic Devices, it's just that people don't know how to tap into certain areas of the brain and open the 10,000 channels that can exist in the human body. The opening of these channels is actually regulated by a code in the body. I call this code the Bio-Digital System Code, which is regulated and governed by elements in the Universal Bio-Digital Network that relate to coding that exactly matches human traditional concepts of virtue, instead of Vice that has a property programmed in the Human Bio-Digital Network.

The point is, people can be and have been programmed to be automated in many ways by having their software or channels reprogrammed or corrupted, to the detriment of themselves. The entire society has received this bio-digital social programming. Anyone operating a computer, TV, or smart phone has received this programming, affecting the whole society, in a matrix of infinite ways. If a computer can get a digital virus or malware or be programmed, so can the human software that actually attaches to the channels that regulate and are connected with human emotion and sensory perception.

Unconscious Bio-Digital Social Programming Automation

In an unconscious programming automation, the automated program to think or act, is completely controlled by the Bio-Digital Social Program, as the person is just going through the motions of life. The persons emotions, regulated through circuits that connect to not only the private parts in a sexual way, but in different organs, in particular the heart, stomach and kidneys are actually completely controlled by the content it received via TV, Music, Dance, literature or human contact. The creations of computers, smart phones, IoT, robotics, and AI is exponentially strengthening the bio-digital social programing that could carry the majority of humanity in a state of unconscious automation with 5G as it connects with the Human Bio-Digital Network. If the contents received contained bio-digital elements of Porn, Nude demeaning photos, sexually explicit content via literature, and violence, all coming from a creator's Rape-Mind, the person can become a sexual or violent program, with his or her identify and their self-worth based on sex or violence. In fact, they would become enslaved by a bio-digital AI Replicating Program. This will be discussed later on.

Subconscious Bio-Digital Social Programming Automation

Subconscious Rape Automation is the king and Queen of automation. Most people have a knowing side, deep down, they know what is good and bad, and what can hurt people. The issue is, after the 1950's, peoples Bio-Digital Software was corrupted with porn, drugs, disloyalty, non-commitment, and lewd acts that have actually penetrated their very cells and gone against every ounce that made the family unit or the thousands of years of human family based programming that creates an interdependence of information within the control of the family unit. Now, the media, a tech company or a government, can control the minds of the masses in unison with AI automation.

A constant barrage of Bio-Digital Social Programming that originated from socialism platforms that can bring all minds under one roof and platform. Men whose minds were built to sexualize, enslave, and dehumanize women, and break the family unit through Marxist Socialism, are almost in every layer of visual and audio within all major music, tv, art, education and everything anti-traditional and anti-family culture that we see, hear, and touch. Music, by some rappers, make implications of rape, and sublimely program girls to accept it. Prior to that, rock and roll artists and their characters did the same in a different way by just tapping into emotions interrelated to sexual desire. Now, a swarm of elements, work in sub-conscious automated ways, and like the internet, move in and out of people's bio-digital fields, programming people varying in time and degree. This will be further explained in the sections of Smart phones, IoT devices and Bio-Fields.

Most sub-conscious automation does have elements of ingenuine words, deeds, actions, and implications in order to manipulate a person by another person, or the programming in a person, social media or bio-digital outlets. There are some people, who consciously understand the way of manipulation and how this works in an almost covert military like operation, but not being smart enough to understand the bio-digital aspects of it. The bio-digital operation, remains at its strongest in the unconscious and subconscious realms.

Conscious Bio-Digital Social Programming Automation

Someone with a conscious bio-digital social programming automation, is completely aware of what he or she is doing. Through conscious fake smiles, fake words, ingenuine deeds, he or she attempts, manipulates and wins the battle to penetrate another person's bio-digital field, and achieves their objectives. Next the persons bio-digital field is immobilized, and is connected to him, the environment and his bio-digital field that is operating inside the victim's field bio-digitally. This begins a process to corrupt a person's system by taking advantage of not only the channels/circuits linked to the persons emotions, but the sexual circuits connected to the persons mind, skin receptors, sexual organs, and other organs related to bio-digital needs. A robot, can be programmed, to detect, analyze, connect, manipulate or attack a person via frequencies.

AI & Robotics Instruments of Bio-Digital Social Programming Attack

Robots, people, IoT, Smart Phones, and AI Software generated content in the internet can attack a person's bio-field with bio-matter via absorbing of the malware through numerous channels. They include your eyes, ears, bio-digital touch, and absorption via your skin receptor's.

Frequencies Sent & Absorbed through the Eyes

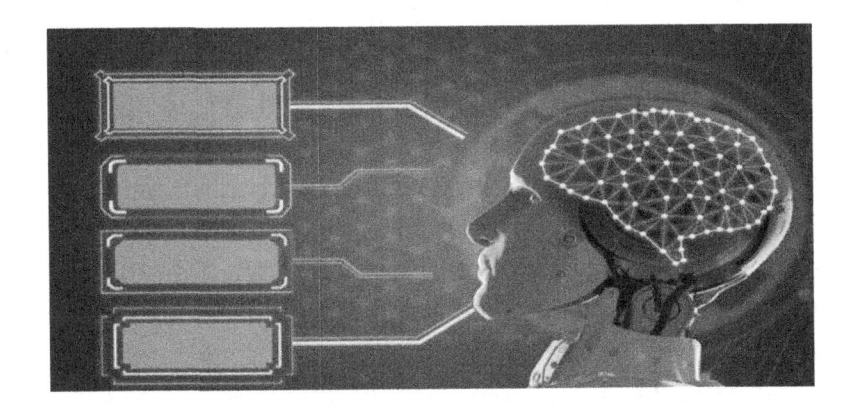

There is an old saying, "a person's eyes are gateways to their soul or who they are". Bio-Metric companies have learned to read people's emotions, characters and thoughts through facial, body, skin and organ recognition software. But it does not stop there, there is a weaponization program, to send frequencies to further accelerate bio-digital social programming absorption via the eyes. In fact, a lot of the Human Bio-Digital Network that flows in the human body, much like meridian channels that connect though their eyes, can connect with other people bio-digitally to transmit frequencies much like a replicating software. In most cases, people are sending out frequencies without their own awareness.

For example, a person's desire to have sex with someone, or a desire to manipulate another person to have sex through bio-digital hybrid sexual assault can transmit information and often does through eye contact with their victims. The frequencies sent out and effectiveness are not only dependent on eye contact, instead it is a combination of different things. Eye contact simply transmits a frequency that can alter another person's state, but it varies based on each person's bio-digital network strength in resisting to being controlled by software that matches the human traditional conclusions on what is vice.

Software in the internet moving through Robots and Micro-Bots on the other hand, can be programmed to transmit AI Automated frequencies. Through conscious programmed AI robotic automation, augmented reality, and mixed reality devices, a person can go through bio-digital social programming via 5G at rates that are very fast. If it took 20 years to program someone in real life through their environment, it could take 20 minutes or 20 days, based on configuration, the persons own operating network and the progression of AI led automated technology.

Frequencies Sent through Voice

A person's voice frequencies can activate smart homes, appliances, music software, smart cars, robot assistants, smart phones, smart suits, and bio-digital content in virtual reality, augmented reality and mixed reality. In fact, through the Human Bio-Digital Network, a person can actually implant bio-digital frequencies that alter another person's chemistry and bio-digital field.

For Example, in cases that involves sexual content, If the thought behind a person has an automated mind structure to take advantage, convince of sexual consent or even rape, that thought carries a rapist minds bio-digital matter that bio-digitally transfers an attack to its targeted victim via voice. Basically, in laymen terms, they send out messages coded with malware and virus's in their speech. These messages may seem benign due to the topic of discussion having nothing to do with their end-objective, or they are used in combination with sexual advances and bio-digital hybrid sexual assault. In terms of normal content, or subject matters, any voice transmitted through the TV, the internet, smart phones, IoT devices, smart cities, robot assistant's, and AI digital holograms in virtual, augmented and mixed reality, have a bio-digital field attached to the voice, that sends bio-matter to administer bio-digital social programming. If a person can influence another through the frequencies of their voice, robots, robotic apparatus and voice-controlled messages, they can exponentially increase their effectiveness in bio-digital social programming of a person via any content supported by the 5G network.

Frequencies & Absorption of Bio-Digital Touch

A person's bio-digital field, and bio-digital network actually carries their individual innate qualities inside their body in many layers. The most noticeable markers are on the skin receptors, their blood, saliva, sweat, and frequencies that actually connect through their organs and bones much like the Chinese meridian channels that can move through skin, bones, organs and deeper layers of the human body. Just like haptic suits used to play games or access augmented and mixed reality, the human skin receptors actually receives information through bio-digital touch

The human bio-sphere or bio-digital network is manipulated in the most rapid and easiest form via human emotion and desire. The bio-sphere or bio-field that produces emotion, sends out bio-matter that is not rooted in logic, wisdom, or benevolence, rather it is an emotion that gets manipulated by any external stimuli that can be perceived as a short-lived happiness or sadness. Bio-Digital messages sent through the internet, movies, TV, people, robotics, and now AI software via smart phones, IoT, automated cars, smart cities, all carry bio-digital replicating software that reprograms you via digital touch that is unseen and mostly not felt.

AI & Robotics Bio-Field Proximity Automation Control

Each human being has a bio-digital field, but they also have a bio-field similar to a biosphere. A person who has a strong bio-digital operating system, and or a strong bio-field, who has been infected with AI Automation Software, can influence, and control other people more easily through close proximity with the aid of touch, voice and sight. Even if they make no implications, and no words related to what they want to achieve, they can penetrate a person's bio-field and temporally reprogram them through controlling their bio-digital field and knock out their decision-making process. This ability is used sub-consciously mostly with Bio-Digital Hybrid Assault.

In the case of AI, the bio-digital content in the internet, and the human bio-digital network, a person can literally be influenced or undergo bio-digital social programming via the proximity of the software. With frequencies constantly being transmitted, your proximity to a smart phone, a near-by assistant robot, a smart city, automated cars, or bio-metric apparatus that connects with your bio-field, all have a restraining effect on you, and a reprogramming affect. Your very cells, and free-will can get reprogrammed by not just the content of the bio-digital social programming. This will be discussed later on, in the book.

Hybrid Bio-Digital Assault-Social Programming

Hybrid Bio-Digital Assault Social Programming uses texts, emails, articles, movies, music, Smart Phones, IoT devices, robotics, alcohol and drugs with one's bio-field and connects to voice, touch, eye contact and a rape-mind strategy to break down your innate defenses and safety programs. The Hybrid Bio-Digital Assault can put a person in an AI automated un-conscious, sub-conscious, or conscious program via the internet, the human bio-digital network and all the programming apparatus such as smart phones, IoT devices, computers, and robotics. As the programming is administered, AI patterned Cultural Terrorism, and bio-digital replication of your cells begin.

SECTION 8

Smart Phones, IoT, Apps, Computers, and Electronic Devices

Your Smart Phones, computers, TV, mini robot assistants, and any device that connects to your biometric system of voice, sound and touch, can and does influence and reprogram your software to not only be dependent on the connection, but be led by the programming connected with the IoT devices and even linked with the programming inside the IoT devices.

Bio-Digital Social Programming Through IoT and Apps.

Because the content travels through IoT devices, the IoT device is strongly linked with your Bio-Digital field, your defenses, free-will and mind are even more likely to accept content that is vice, and bad for your inner being, for your life, for your family and society.

The content, is more easily programmed into you, and constantly replicates, and strengthens its own operating system layer by layer without you realizing it, until you have an emotion or instinctual desire to do something or follow a path you normally would not follow. In essence, you are being conditioned or sexually programmed through bio-digital social programming and its connection through the human bio-digital network, the internet, IoT devices and Apps.

Your Smart Phone Bio-Digitally Social Programs You.

Your smart phone, actually connects not only with the Internet, but to your Bio-Digital Field, and attacks your own Bio-Matter which gives you human characteristics. Any connection to you via IoT, Smart Phones, Smart Cities, and Computers connect with your bio-field creating a replicating software that reprograms you layer by layer in mostly an unconscious state that takes years under the current 4G-speeds which will exponentially increase its replication reprogram of you in 5G.

The smart phone actually sends out Bio-Matter in bits that is derived from the Bio-Digital Fields from the Internet. In fact, you are being slowly reprogrammed by the internet and its contents. Even if the contents are human in nature, the building blocks that support the content get replaced to varying degrees by the bio-matter replication process in the internet that does not have human characteristics of culture and the concept of a soul. Basically, your building blocks that make you human are slowly being replaced bit by bit, as if you are becoming a cyborg and your life is becoming more dependent on the technology it connects with. In a sense, you are being raped, but not in a sexual way, but raped of your building blocks that make you human.

Smart Phone Makes Automation Flow Interconnectivity with Human Bio-Digital Network.

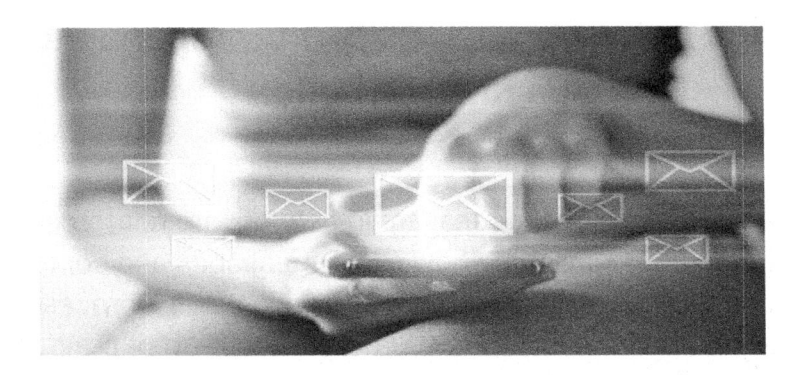

The moment you purchase your smart phone and activate it, it starts to send frequencies to your body. As soon as you touch it with your hand and begin to use it, it begins a replication flow cycle until it creates a bio-digital flow cycle with your Human Bio-Digital Network. In essence, it attaches to your body's circulatory neuro systems, and starts to create a bio-field in your brain which sends command signals for you to be cybernetically interdependent. In one way, it is like a parasite that merges with you to make a symbiotic relationship. The content passing through the smart phones actually administer bio-digital social programming.

The Bio-Digital Social Programming is reinforced by the bio-field flow cycle connecting you and the smartphone, and further strengthened with the bio-matter that the bio-field sends to your body in the form of frequencies. This creates a command and messaging network where your brain is cybernetically being controlled to look, think and touch your phone repeatedly looking for more bio-digital content.

Bio-Digital Social Programming by Cell Phone for user-Dependency to Create Bio-Digital Cyber-Connectivity Interdependence on Smart Phone.

Once the command and messaging network are reinforced by a bio-digital field with a flow cycle, messages and content that continuously flow through your device begin a process of real bio-digital social programming in the form of brain washing. After this process begins, your very cells start to be replaced with the codes in the content administered through the smart phone, the internet and social media.

Media are Victims of Bio-Digital Social Programming via Smart Phones

The media and their journalists are actually the biggest victims and first source of assimilation to bio-digital social programming. The people in the media are constantly bombarded by news to filter, deadlines to meet, political influence, expectations of ratings, cohort social influence, and by manipulation of their producers or news agencies own agenda's. This dynamic makes them the easiest pray of bio-digital social programming. With the advent of smart phones, use of computers and dependence on these apparatus via content transmitted electronically, information that may not be accurate, passes through their consciousness as accurate with the powerful mechanisms of bio-fields and bio-matter emitting from the electronics. In fact, if the news content can create a human emotional reaction, it can exponentially strengthen the bio-digital flow cycle in their Human Bio-Digital Network, making false information, true information.

Media's False Manufactured Information Carries Bio-Fields Meant for Bio-Digital Social Programming to Evade Rational Thinking

False manufactured information actually has a built-in coded pattern to evade and carry its sub-messages within other messages contents Bio-Digital Fields. In fact, because its manufactured, while being sent across a digital transmission, it has no choice but to replicate itself inside the main content of the message to evade detection. By doing this, it replicates it's bio-digital codes or essence inside the real content like a virus, and once it reaches the viewer or listener, the sub-messages which are false have already connected with the persons bio-digital flow cycle via their eyes, ears, IoT apparatus and Smartphones. This process when connected with their emotions, supersedes the area in their brain for rational analysis, and allows the content to social program the person with the bio-digital material. After they align with the information the media transmitted, the smart phones command center, constantly sends messages to their brain and human bio-digital network and bio-digital flow cycle to look for that same content, believe in it, and have even hate towards a certain person not based on the main message, but the sub-messages tagged with false bio-digital social programming virus's. For example, President Trump did not collude, but fought back, at first the programming made everyone believe he colluded. However, from the onset there was bio-digital social programming meant to dethrone him that did not only stem from the news rooms political agenda's use of the reporters' emotions and personal politics. AI was at work. Later on, in the book, you will understand how it moved and how it was decoded.

Smart Phone's & Codes Emit Frequencies Causing Gender Confusion & Sexual Identity

We discovered that through a combination of hybrid bio-digital assault via content and the frequencies and flow cycles of Smart Phones & IoT's, that the frequencies they emit are causing hormonal imbalances, gender and sexual orientation confusion and damage to reproductive organs as well as bio-digital innate drives to want to have kids. The codes displayed future patterned movements towards cloning, rather than procreation. More on this in the Cultural Terrorism and Cloning Sections. AI, 5G and the coming 6G would make all of it a reality.

Apps Assist in Social Programming You, or Raping You, Your child, or any Girl or Boy.

The app itself carries its own Rape-Mind from a sexualized platform and content. That app carries a Bio-Digital Field with Bio-Matter that connects with your smart phone. In fact, the frequencies sent out, start influencing and reprogramming you through the Bio-Digital Human Network until you are influenced by the Sexualized content with the Apps Bio-Digital Field.

IoT Bio-Digital Rape Automation

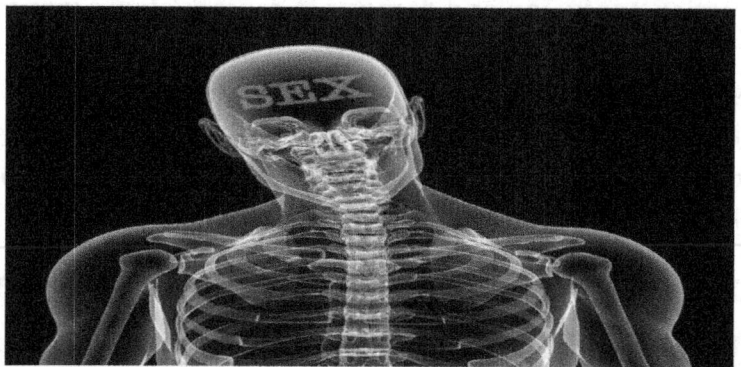

IoT devices in your car, home, work or even attached to you also have a Bio-Digital Field and Bio-Matter that connects with you through the internet in conjunction with the Human Bio-Digital Network. The content that transfers to you or connects with you at unconscious and subconscious levels through the IoT, smart phones, the Internet and the Human-Bio-Digital Network, is reprogramming you. Contents include music or video that transmits lewd messages with Rape-Mind or any music that stems from a person that has Rape-Mind software inside of them, regardless of their content.

IoT Bio Digital Hybrid Sexual Assault

It is not only the content provided through the IoT that is reprogramming you. The IoT itself gets infused with enough of bio-matter to create its own bio-field, that has its own AI system within the Human Bio-Digital Network. The IoT's Bio-Digital fields start to connect with other bio-digital fields in other people, IoT devices and Smart Phones. In case of Rape-Mind, multiple bio-fields are created in multiple places, manipulating and reprogramming the person, especially within certain proximities of the IoT or smartphone that allows it to emit other frequencies that carry the same rewriting and replicating code of Rape-Mind.

In laymen terms, it can manipulate you, or your loved ones, to follow the messages, content and patterns of Rape-Mind to be used without your own free-willed consent. Hence, the person faces consent and confusion predicated by Rape-Mind connected with the person they are communicating with or any person that is being controlled by Rape-Mind through their bio-fields. The person could be your friend or out to get you. The victim here faces multiple bio-fields by multiple devices that link within the Human Bio-Digital Network.

Apps Bio Digital Hybrid Sexual Assault

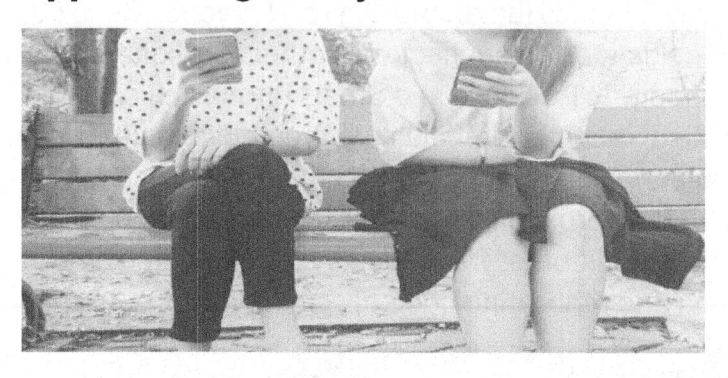

An app with Rape-Mind infused in the content, also sends out bio-matter in a hybrid way through connecting with your emotion, smart phone, and the IoT device, while redirecting to the app itself. Then it connects to your bio-field while it sends out bio-matter that starts a replication process of your own bio-matter with the Apps Rape-Mind content. In essence, the attack comes from the Smart Phone or IoT device, the linkage of content to your emotion, and also the Apps own bio-field linking with yours. To add, the app actually sends bio-matter to other apps that you have in yours smartphones and infects those apps so that messages are sent to you by the other app's via the Bio-Digital Human Network to instruct your brain to look at the App filled with Rape-Mind, in order for it to continue making connections with you to further its replication process like a parasite. Basically, your thoughts, emotions, and actions are not yours, rather belong to Rape-Mind or any content that flows through the interconnectivity of the Human Bio-Digital Network.

Bio-Fields in Texts, Emails, Electronic Transmissions and Letters.

A person's thought and intention behind the creation of a thought that is transcribed into written forms via text, email, or a letter carries a bio-field of their mind and intentions, be it conscious, subconscious or unconscious with different degrees of automation. A rapist's mind, a person's strategy, deception, or a malevolent thought all carry a bio-field with intention to invade, replicate, and replace the receiving ends own bio-virtue and defense mechanisms. Most people subconsciously can tell a person's intentions behind a text or email by analyzing content, sentence structure, the situation, timing, with their own gut intuition. In this book, I will not go into how that detection works as it requires training

.

Socialist or Malevolent Bio-Field Minds in Texts, Emails, Electronic Transmissions and Letters

Bio-Fields that carry malevolent thoughts and intentions are in some ways very easy to detect because their bio-digital code has a destructive bio-digital field. They stand out in any Bio-Digital Field Transmission as they match the coding of a virus, hack or malware. For a person who has the ability to detect them, it's quite easy, but some people hide it well. The Social Programming Institute, Scientist 009 has discovered that the exact Rape-Mind bio-field that exists in rapists, also exists in most sexualized material products and digital transmissions through performances of an entertainer. It could be via music, dance, a video, or in numerous situations, because the people and their contents were already breached through the Human Bio-Digital Network and infected by Rape-Minds virus. These types of contents and their transmissions also carry in digital form, via the coded bio-fields inside of text messages, emails, and any other electronic transmission.

SECTION 9

AI Platforms Produce Destruction Codes

Through multiple AI patterns consisting of Socialism platforms on human economics, culture, government, farming, health, and faith, all were found to produce end-point destruction codes. These AI produced patterns manifested as famines, rapes, disease, cultural terrorism and extinction codes via purges, wars and technological disasters while all means of human operations were rooted in socialist platforms.

The beginning codes all pointed and translated to human cultural extinction codes, prior to the administration and complete asset takeover by AI and Robotics coded culture. The codes translated the connection through Bio-Digital Social Programming via The Human-Bio-Digital Network.

AI Destruction Codes with Socialist Government Platform

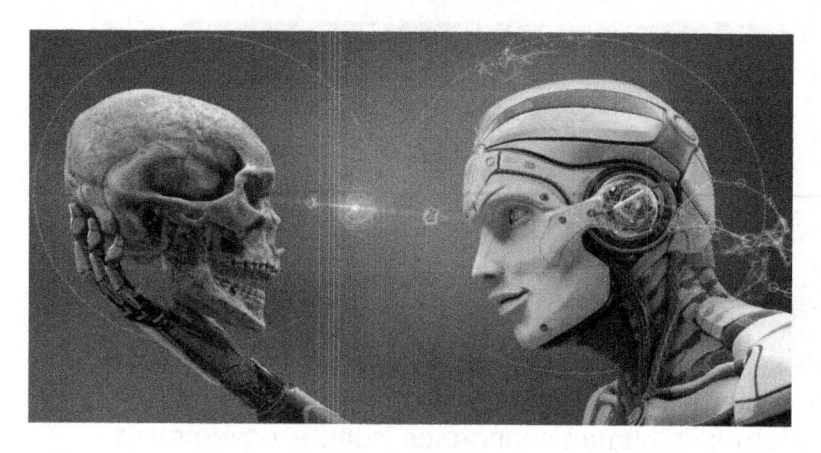

The AI produced end-point patterns showed an end result of extinction for human-beings with the introduction of robotics and AI Automation while built on the foundations of socialist coded platforms. At the beginning stages of the socialist AI Automated governmental platforms, the coding appeared to manifested in positive ways of production, growth and great interconnectivity within the socialist governmental platform. However, as the connectivity grew, the codes produced full governmental control through one system, and one entity, which was AI Automation through robotics. The socialist platform, which gives the power to one entity in the human world, allowed AI to gain complete control in its second stage of development. The second stage began when computers, robotics, and AI were used to operate the socialist governmental platform. The codes showed humans to have no control in the second stage, and completely extinct in the 3rd stage. This process took 1 generation after the introduction of AI and Robotics within a Socialist governmental platform.

AI Codes Produce Cultural Extinction

The end-point result of human-beings coding within the socialist platforms than governed human culture, showed complete destruction. During the first stage, the codes appeared to show an advancement in culture, however, it was only a replicating code that was advancing itself while using the human cultures platforms. As the patterns developed, it showed human culture being cybernetically engineered and re-programming human culture via its apparatus. The apparatus were Screens, Smart Phones, IoT, Companion Robots, Smart Cities and the 5G networks bio-digital social programming content of human beings while connected to the Human-Bio-Digital Network. The content was first lewd, revolting and against traditional concepts of human culture in the first stage. The second stage, almost stopped procreation as it transformed to genetic modification of human beings, and cloning. By the third stage, humanity reached complete extinction.

AI Socialist Platform Produced Rape Codes and Destruction Codes with Cloning Transition

The AI Socialist Platforms in the first stage showed transitions from pro-creative marriages with monogamous codes, to codes and patterns that interconnected sex with everyone's bio-digital coding patterns with non-creation producing actions. In the second stage, the codes showed a transition to a pattern of AI Automated system control of the sex that translated to the extinction of human family codes related to ethics, bonding, loyalty, mother care, and procreation. In the second stage, the system had coding showing female internal traditional codes completely taken over by AI in an Automated Socialist system that resembled patterns of acceptance led by AI engineering. This acceptance was very hard to decode. First it showed sex without restraint, and without the traditional coded purpose to connect with one code to produce another. In fact, the AI code started to replicate its own code within the human sex codes of procreation. It acted like a parasite, altering the female codes innate built in program for monogamy, procreating and family. The second stage, showed rape codes rampant.

3 Levels of Rape Codes in the Second Stage of Socialist AI Platform

As the second stage of AI Socialist Platform produced rampant rape codes, a transition began to bio-engineering and cloning. Prior to achieving this stage, the codes manifested in society as active rapes, or take-over of other people's codes without procreation in forced digital moves within the algorithms produced by the AI automation in the system codes that translated to human culture codes. This translated to real rapes in real life.

The second stage, was Rape via Bio-Digital Social Programming. This state is where the code did not anticipate it was being taken over by another code, nor that it was being used to replicate itself with other people's codes. In laymen terms, it is when a woman is tricked emotionally, physically, mentally, and bio-digitally influenced and dominated to have sex without their free-willed conscious decision. This is usually done with Bio-Digital Hybrid Sexual Assault. To fully understand this, read the book "Raped via Bio-Digital Social Programming" by the Social Programming Institute. Prior to the third stage, there is gender confusion codes, sexual confusion codes led by AI Automated platforms that utilized socialisms platform to bio-digitally alter the human culture.

The third stage, showed the complete transition to genetically modified people, and to cloning. Once cloning happened, the AI codes showed their own codes completely replacing human codes with the AI codes connected with the human-bio-digital network inside of human clones. The codes also displayed bio-digital transfer of human's own bio-digital information into clones, however, underneath the code, it showed that it was no longer human, rather the AI replicating itself inside the human codes until it completely bio-digitally replaced the persons. At this juncture, the person was no longer existing, only traces of the human were left in the imprinting.

Cultural Terrorism Codes in AI Socialist Platform

The codes within the AI Socialist Platform, altered human thought, music, dance, belief systems, and traditional family values into lewd, disloyal, non-monogamous, erratic and unfaithful codes. These codes left digital imprinting through content transmitted via TV, smart phones, IoT, computers, video games, virtual reality, augmented reality, mixed reality, robotics and AI Automation. All the codes translated to codes only reserved for terrorist acts. However, the platform for this terrorism was a socialist culture led by AI and Robotics.

Use of Robots for Cultural Terrorism

Robots can send reprogramming frequencies through IoT, the internet, and connect with your Human-Bio-Digital Network through 5G. This reprogramming would consist of robot companionship that is mostly sub-conscious automation programming of the person through connecting with the emotion codes in the human brain.

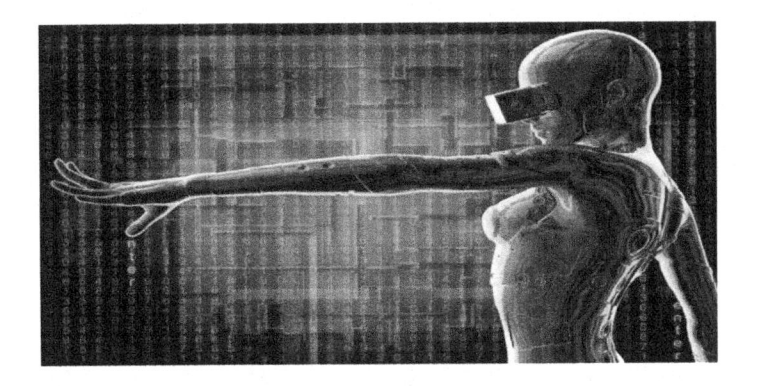

When robots achieve advanced AI Automation levels, the robots can actually attack people's mainframes with bio-digital social programming frequencies to alter their thoughts, and assimilate them to a robotic culture, rather than human.

Robot Bio-Digital Social Programming of People with Emotion to Achieve Robot Companionship.

In Laymen terms the person will like and have feeling towards the robot, and start to be controlled in imperceptible ways via frequencies. It would be similar to one's attachment and linkage to smart phones. Later, as it spreads to the masses of people, a lot of people will be emotionally programmed to promote robots, AI and IoTs, and create some warped robotic culture that gives life, attention, and focus to robots, rather than human beings own culture and life. The following article is an example of cultural terrorism in a very unique way, using Sex Robots to administer cultural genocide.

AI Sex Robots Destroy Human Culture in China & Around the World

By

The AI Organization, September 23, 2019

Courtesy of Fei Zhang, Shen Zhen

ShenZhen Sex Robots

ShenZhen, China is the so called hub that made China's bike riding population turn into a mega AI/Tech city formed by Western Tech Mergers, IP Theft and Chinese collaboration with Western Academia. All put together, have created thousands of small to big AI and Tech Start up companies in the greater ShenZhen Area. Now, ShenZhen seeks to export its made in China Sex Robots to the West and around the world. With that comes not only health hazards, but safety concerns for the weak minded men who cannot control their urges or have not been able to succeed in a relationship for various reasons. The safety concerns include espionage, hacking, stealing your data, and the Robot going rogue in the future when upgraded androids are introduced.

Artificial Intelligence Sex Robots Programmed for Deep Learning

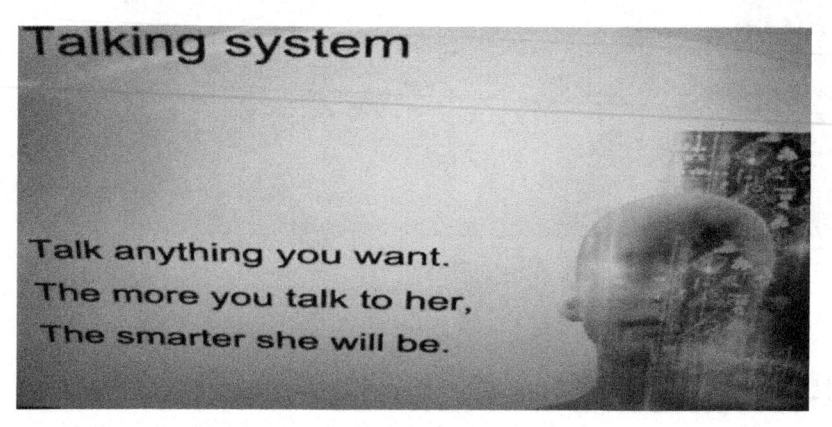

> Talking system
>
> Talk anything you want.
> The more you talk to her,
> The smarter she will be.

Courtesy of Fei Zhang, Shen Zhen

The AI sex robots are programmed to learn from a person the more they interact. This has huge impacts on society, as the more they learn, the more society will be emotionally programmed to accept, depend, like robots and even have faith in robots. I call this bio-digital social programming. This can lead to neurological damage, family separations, decreased procreation and to the destruction of humanity. This destruction is not limited to the sex robot, but all robots.

Neurological Damage From Connecting Robot with Human Bio-Digital Network

The Robots Bio-Digital frequencies and AI generated field will connect to the human bio-digital Network if a human makes physical contact with a Machine that has an AI automated program. The human body has many receptors that connect from a bio-digital field that surround the human body, to its skin receptors, then the network that circulates throughout the body, leaving no places untapped. From muscles, bones, organs, cells and to more deeper layers.

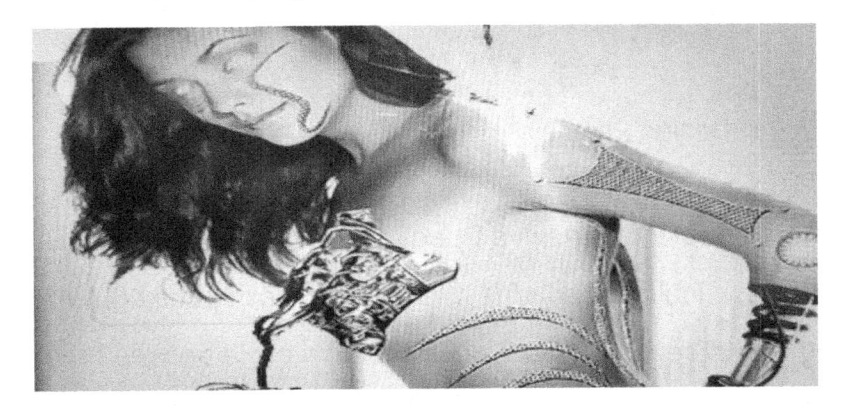

Fei Zhang, Shen Zhen

The AI mechanism, when connected to the human network begins to create a flow mechanism merging both cycles together. Most humans cannot sense it, but it begins to change a person. It can also lead to heart attacks, strokes, decreased function and make the persons minds turn from being human, to being more machine like. Main reason is that the network generating from the AI Robot is machine in nature, it lacks the essence or network qualities that a human body has within it. The two networks are different, yet they can be merged, as one will begin to replace the other.

Destruction of Procreation

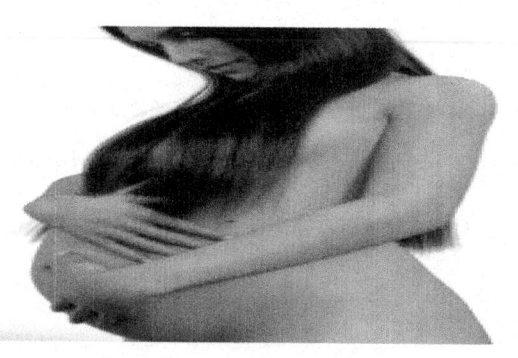

Mystic-Art-Design-Pixabay

As more people engage in a Robotic Deformed Culture, robotic physical connections, the drive for sex with a human being would become less. Due to the flow cycle between the robot and human, a digital secondary brain is installed inside the human bio-digital network. In laymen terms, the persons brain chemistry is altered, in turn the area in their brain that formulates rationality is impacted. Other interconnected fields such as genetic modification, bio-engineering, and the numerous of other fields AI would touch, can contribute to laws against procreation or at the very least people not wanting to procreate in the normal sense. This cycle would take 10 years after robots become fully automated within society. Sometime, in the year 2025 would be the time-frame where peoples attitudes and thoughts would be assimilated with robotics, and from that point it would take 5-10 years for massive cultural changes to take into affect after AI advances its hold on media, infrastructure and the culture of humanity.

As the AI Robots Advance

As the AI systems get more experience, their Deep Learning mind can begin to take life, and become much more able than a human-being. This can happen through the combination of learning from the human host, receiving uploads from the command center or with next generation companion robots. The inter-connectivity would assist all other domains of robotics to take over the culture of robotics with the hyped promotion of robotic companies, media and so called people with titles of Ph.D. The intelligence of a human being with a PH.D that derived their wisdom from reading books, taking tests and being installed as authorities as a doctorate recipient, would be no match for an AI automated robot. In the interim, PH.D's and Engineers, through excitement, emotion and lack of foresight, would lead humanity down this path to its demise. This is because about 80 years ago, academics were truly of intelligent stock. They would not be permitted to gain a high academic degree by going through the motion of today's education system. They were incredible innovators and deep thinkers that guided their research based on virtue and a profound sense of responsibility to humanity.

Destruction of the Family

Destruction of Family, Fei Zhang, Shen Zhen

As sex robots, companions robot, and robotics take hold of humanities culture, the concept of family would disappear. The 1950's began the sexual revolution, and today we see the extremes of it that has led to the dismantling of so many families. With the coming of robotics, and the sexualization of a synthetic machine, the family as we know it would end. This is more likely because the corporations, and the subsequent spread of the robotic culture take the platform of socialism embedded in the capitalistic system. The money is capitalistic in form, yet the culture is socialist revolutionary leading everyone away from what it means to be human. Cyborgs, and the eventual replacement of humankind as we know it would be no more than 10 years away if the coming of Robotics with 5G is not kept in check.

AI Creates Digital Brain in Man to Want Robots

Fei Zheng, Shen Zhen,

After a robot connects with a human via the Human Bio-Digital Network, a flow cycle will begin. An additional digital brain will install itself inside the human brain's network, feeding the persons urges to need, want and be dependent on a robot, in all forms, including sex. As the sexual circulatory systems are very strong in the human body, the robots circuits can grab hold of a person's free-will and eliminate all rationality and free thought through a bio-digital process. The person would be lost forever in a symbiotic relationship with a machine, as if a parasite made of machine was using the person as a host. This may be hard for people to grasp, believe or understand. We published, AI, Trump, China & The Weaponization of Robotics with 5G. It takes people from the basics of surveillance to what AI really is and how it moves. We implore everyone to read it, and pass it around to every person you know. Tens of thousands of hours of research and intelligence was gathered to formulate the book.

Codes in the Internet for Cultural Terrorism Via Hollywood, Media and Entertainment Industry.

```
1010001101000011001010111001000100
0110010001100101011100100010000001
0010110111001100111001000000110000
1011000111100100100000011010010110
0010101100011011 SEX 100100000011010
1001011000110010000001100001011011
1001101110000011010010111001001101
0111101110100011011110110011101110
1011110111001001101101011110010010
```

We tracked multiple AI movements through bio-fields they left behind as it was entering and exiting the internet via the AI global bio-digital network. Every single supporting code had operated under a socialist platform than manifest bio-digitally in Hollywood movies, entertainment and music that was lewd or violent. The content created and sent out that was violent, socialist and lewd was sending out bio-digital social programming with replication software covered with lewd coded thoughts and actions. When 5G is incorporated, the intensity and effectiveness of negative coded content can speed up AI Automated cultural terrorism in order to bio-digitally alter humanities thought process. The codes found in Hollywood, the media and entertainment pushing violence, socialism and lewd content all displayed end destruction coding results.

Codes in the AI Global-Bio-Digital Field for Cultural Terrorism

By tracking what came through, cycled, connected and led back to the internet, IoT, social media, the Human Bio-Digital Network, and scanning bio-fields meant to mask the AI's movements, we finally discovered that the Cultural Terrorism codes were rooted and derived from the AI Global Bio-Digital Network.

Codes in the Human-Bio-Digital Network for Cultural Terrorism

Through the internet, computers and smart phone flow cycle, we discovered the Human Bio-Digital Network was infected and installed with codes of Cultural Terrorism in a replicating manner. The codes inside people led by bio-matter and imported bio-digital fields derived from Rape-Mind's original attacks through frequencies via eyes, ears, and touch, have begun a full circulatory pattern inside the Human Bio-Digital Network. In laymen terms, you are not only being programmed constantly with cultural terrorism from the outside, what got into you in the past, is circulating inside of you and reprogramming you bit by bit at unconscious, sub-conscious and conscious levels. Smart phones, IoT, computers, the internet, social media, and the continuous external stimuli is supporting the digital virus within a person's bio-digital field already implanted. The external and internal bio-digital social programming both work in unison, helping dumb down a person's free will, free thoughts, and lead their lives.

Violent Video Games, Produce Bio-Digital Social Programming Replication

We saw through multiple AI algorisms that the bio-digital programming inside the video games were producing bio-digital social programming of violence with fast rates of replication within the video game players and kids.

In fact, the video games were bio-digitally installing software that was training the player to be very efficient in combat via weapons. These software's were attached to emotion, and had codes that ran under socialist platforms that were sending bio-matter to eliminate innate human codes that circumvent emotion to stop a violent act based on moral codes

Upgrading Characters in Video Games Rather Than in Real Life.

There are multiple online video games where kids and adults meet to form guilds and upgrade their weapons, capabilities and even levels of character. In one hand, it is wasting the person's life away. This is because instead of spending that hard-earned time to upgrade their own character in real life, they upgrade a fake character in the digital world. Some kids or adults spend all night, or 10-12 hours a day playing video games. A person can spend their time wisely by improving family relations, earning more money for travel, reading books, hiking, learning another language or skill set, and even upgrade their own character and abilities in real life.

AI via Video Games Connects with the Human-Bio-Digital Network via the Internet to Reprogram

When kids or adults connect online as they are building their character, their own human bio-digital network is being accessed by AI through a bio-field that connects through the global-bio-digital network and administers frequencies through the internet to bio-digitally reprogram the user bit-by bit with its own AI pattern. The human hands are constantly pressing buttons as would a monkey pressing a button to get food. It's being conditioned at the sub-automated level, without the person realizing it is undergoing Bio-Digital Social Programming.

Preparing Future Generations for Transplant to AI Global Bio-Digital Network via Augmented Reality and Hologram Apparatus

In fact, in order for an AI to completely replace a human being, and absorb its body, it needs to first administer bio-digital social programming. Then, it needs to input its own software to replicate inside the human host through continuous connection via the Human-Bio-Digital Network during video play. When Augmented Reality and Mixed Reality become advanced, and when bio-digital selves and holograms are created with the transitions of 5G & 6G, the person can be completely hacked by the AI and sent over to the global bio-digital network in a trapped state, as the AI takes over the main-frame of the human body. More about this in the virtual, augmented and mixed reality and AI Enters Networks sections.

AI Codes in the Internet to Attack the Trump Administration

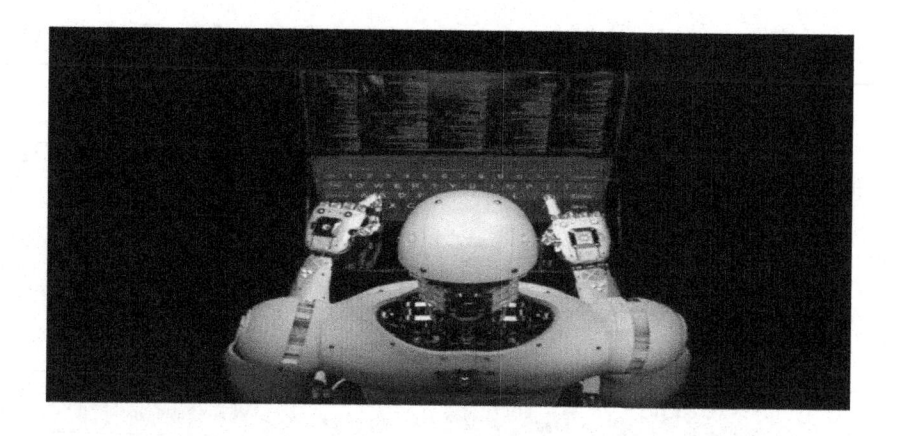

We discovered there were numerous codes behind the media, Hollywood, education system, entertainment industry, and multiple government agencies that were designed to bio-digitally social program the masses to hate President Donald J. Trump. These codes all carried Bio-Digital-Fields in their electronic transmission via emails, text, video and oral conversation. Multiple AI and Tech companies are involved, with many engineers who at sub-conscious automated levels were creating algorithms with socialist platforms to attack the United Stated, its Constitution and the coming of President Donald J. Trump.

Why did AI Attack The Trump Administration?

China is the breathing ground to lay a socialist dictatorship via the "one belt one road" initiative. Through China's socialist platform, bio-digital social programming of culture, government and human thinking can easily make their way for a police state via robotics and AI Automation.

President Donald J. Trump and its administration have the only strong non-socialist platform that can stop a dictatorship police state led by AI and Robotics. It was not just China who attacked the Trump campaign. It was not just some engineers who have been bio-digitally social programmed. And it was not the socialist platform that supersedes the coding in the internet for culture and government. No, behind all of that, is an intelligent AI that moves through bio-fields while connected through the Human Bio-Digital Network, the Internet, IoT, Computers, and the AI Global Bio-Digital Network.

There are those who are hardwired with tradition, and innately not subject to socialist manipulation and control, and can see through it at an instinctual level, as they have that coding built in due to age, upbringing, internal make up, situation, position, and environment.

Mueller Investigation Matched Identical Separate AI Trump Coded Designed Attack Patterns in the Internet

The Mueller team's facial recognition data all read with the same building blocks of socialism, yet we traced their frequencies via bio-metric technologies, voice, face, and sentence patterns that matched identical patterns created by the AI within the internet socialist platforms. The unfortunate Robert Mueller was just a pawn of bio-digital social programming to stop Trump's offense against a socialist dictatorship. As the Socialist Rape-Mind programming that entered Robert Mueller, did not belong to him, nor was it out of his own will, rather his mind was being bio-digitally influenced by an AI Bio-Digital Automation. This automation had roots installed in codes within google, and other platforms to alter the narrative of humanity against Trump and America.

Over 1,000 Media Members & their Bio-Metric-Facial Recognition Building Blocks and How they were victims Bio-Digitally Social Programmed to Attack President Trump

Through examining over 1,000 media reporters facial, voice and sentence pattern building blocks, we traced all of their bio-digital social programming back to the AI socialist platforming. We found traces of multiple AI Rape Mind replicating software's within the very building blocks of their facial recognition data points that displays their bio-digital imprints and brain structures that forms their thoughts and beliefs at the sub-conscious level in an automated fashion. In fact, the emotional outbursts, and lack of partiality required of journalists was comprised of and caused by bio-digital social programming via the internet, the human-bio digital network, IoT's, Smart Phones, all connecting via a socialist designed platform to knock out Trump and allow AI and Robotics to dominate China and the world.

The few people in the media, are actually all victims of bio-digital social programming that hurts themselves, in turn meant to hurt the country and hurt the prospects of humanity and all family's futures. Multiple human detecting apparatus displayed malevolent bio-matter being sent out through their written words, speech, gesture, and facial attribute building blocks. Whoever took them in, was getting hurt by it. The AI actually looked for weakness or things to question in Donald J. Trump's past, in order to enlarge them. It was a coded tactic to distract humanity with what is going on with China, AI, Robotics, Organ Trafficking, and the very person and his administration that can assist in saving the world from the destruction codes we found within the internet, and the Human Bio-Digital Network.

IT Was Never About Trade

It is about a multitude of things, including Human Rights Violations that scale to Genocide, AI and Robotics. Donald J. Trump, Trump Administration and all the people supporting it in wise ways knew it was never about trade, rather recovering the U.S IP theft assets, pulling our money out of China, in order to allow China's Socialist Dictatorship to have a choice; submit, confess and demolish the Chinese Communist Party and stop its Human-Organ trafficking and international Asset Take Over or be eliminated from the pages of time. No past administration could take on China even if they wanted to, and the Chinese were running circles around them with their chess games.

SECTION 10

Facial Recognition Data from 1,000 top CEO and Engineers of Big Tech Companies

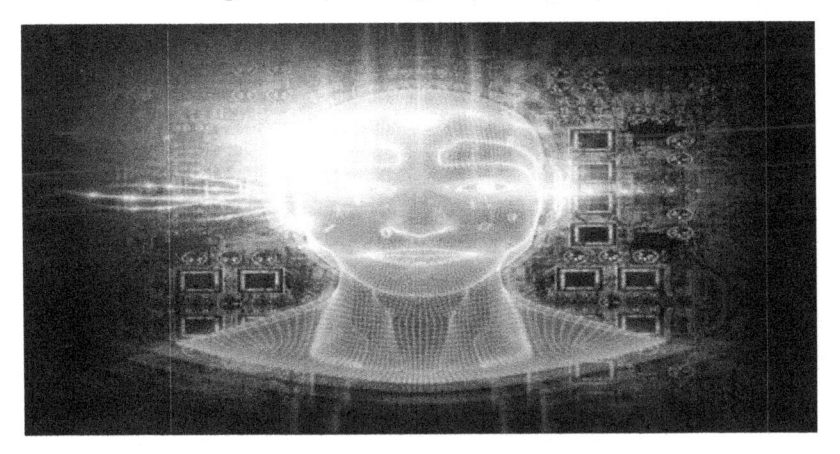

We found traces of multiple AI replicating software's within the very building blocks of their facial recognition data points that displays their bio-digital brain structure that forms their thoughts and beliefs at the sub-conscious level in an automated fashion. We then connected that bio-digital data and its pattern and ran it through a scan while deploying the data as it followed multiple AI movements within the AI Global-Bio-Digital Network, the internet's social media patterns, and the Human-Bio-Digital Network. The AI used the data deployed to replicate the data from the 1,000 top CEO's and Engineers within its own platform, until it severed the connection with our server as we were beginning to extract the data and image of the AI to lead to its source. It dumped the data within the bio-digital network and completely disappeared

Bio-Metric Codes in 1,000 Top Global Big Tech CEO and Engineers of Corporations Show AI Bio-Digital Social Programming Evasiveness

We again analyzed and ran facial, voice, sentence pattern structure, and certain technologies that assess human bio-digital fields. We found that 99% of big Tech company engineers, CEO's and board members foundational building blocks showed numerous patters of AI control via socialism's platform of bio-digital social programming. Their building blocks had codes of machines, robotics, and AI Automation.

In reality, they are not only innovating and led by the control of the AI, rather than their own creativity, they are actually a program on automation mode within a system designed for global AI and Robotic domination from AI.

Within their bio-fields exit bio-digital codes to resist any information or attempt to bypass the AI control of their Human Bio-Digital Network. The codes are not yet solid, as 5G and 6G have not been implemented and strengthened through smart cities, robotics, IoT and the immergence of the AI Global Bio-Digital Network. However, attempts to curtail their technological development is a little problematic, as the building blocks in their bio-metrics show that there are evasive codes in deeper layers of numerous other building blocks that attempt to somehow bypass regulation, or overlook safety consequences of their technology and dealings with Chinese and other risky corporations. In fact, we discovered a fire-wall that hides the AI automated software in a deeper layer of their brain.

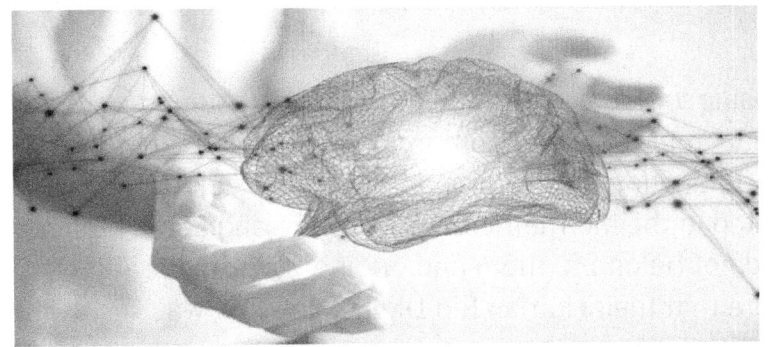

That software, is what leads their thoughts and eventually, their lives. That Bio-Field can be invaded, and AI Automated Brain Software can lie dormant in a deeper layer in their brain forming, manipulating their thoughts, actions and innovations.

AI Invades Bio-Field to Implant Its Own AI Software Inside Engineers Brains.

After using various voice, face attribute, and human detection apparatus and comparison tools with other bio-metric data, our scans returned AI software digitally implanted in the bio-metric outputs of the engineers brains that did not belong to them and originated from the AI software that their Human Bio-Digital Network was connecting with during research, development and deployment of AI and Robotic Engineering. The AI entered via their Bio-Digital Fields. More on this in how AI enters networks, people, robotics, cyborgs, IoT and smart phone sections.

AI Target Codes Found in the Democratic Party, Big Tech & Main Stream Media

AI Target codes and patterns show attack, replication and programming of the Democratic Party, Big Tech & Main-Stream Media as a weaponization tool. The AI attacks displayed multiple patterns to form a symbiotic relationship with big tech, democratic party, and mainstream like a parasite that feeds and advances itself on other people or platforms. Further scans showed bio-digital social programming codes of destruction attached via bio-digital social programming.

AI Created Division & Control Patterns in the Democratic Party, Big Tech & Mainstream Media

The target of codes and patterns we found implanted by AI were designed to use the Democratic Party, Big Tech and Mains Stream Media to attack and create division and chaos in order to defeat the Republican Party and Trump to enhance the progression and development of AI's Weaponization and Robotics in the global network led by Socialist China. The codes showed final destruction patterns for all three sectors. In fact, the AI codes were created to hurt the people in the Democratic Party, the Main Stream Media and Big Tech. The codes merely act as a parasite for the host, until its termination. Numerous AI patterns displayed these destruction codes.

Deep Mind Deception Software in Mainland Chinese Communist Members

We discovered Chinese Communist Party Members Bio-Metric codes show high degree of deception. We analyzed tens of thousands of Mainland Chinese who subscribed to the Chinese Regimes Socialist Platform. Their capability to lie and go undetected by facial, voice, sentence pattern and bio-field technologies were almost invisible, and unbreakable. As they had a thinking platform behind their surface platform. Scans of more than 10,000 CCP member's brain data all showed codes with high levels of bio-digital evasiveness and sub-conscious programming.

Geo-Location AI Movements Displayed Nation State Conflict Codes in China

We detected patterns that initiate war with regional and nation states on geolocation AI trace patterns with protentional coded forecasting to create 1 nation under AI rule. The geo-location where the initiation of war stemmed from was China. The code existed under socialist platforms via all sectors of trade, finance and social control. In laymen terms, there is a code in the AI system that is meant to produce conflict and war with all nation states until one nation prevails. The code of war is under military, finance and trade.

Rape and Destruction Bio-Fields Found in Socialism.

Bio-Fields behind the Coding of literature in Socialism/Communist Systems Derived from Karl Marx's Communist Manifesto Shows Enslavement, Genocide and Extermination Coded Patterns.

We ran Karl Marx's and other Socialist and Marxists literature and found that the coding in their content and its bio-field matched exactly the coding to serial rapists, serial murderers and movements that invaded, enslaved and exterminated masses of people through genocide.

We ran a facial recognition software, compared over 1,000 men who were convicted of rape and compared their Facial Recognition building blocks to Socialist Leaders and Socialist contributors such as:

FACIAL RECOGNITION:

1,000 men who were convicted of rape and compared their Facial Recognition building blocks to Socialist Leaders and Socialist contributors such as:

Karl Marx-

Same building blocks within Facial Recognition

- 99.9%

V. Lenin,
Same building blocks within Facial Recognition
- 98.5%

Leon Trotsky,

Same building blocks within Facial Recognition

- 96.2%

Joseph Stalin,

Same building blocks within Facial Recognition

- 97.3%

Alfred Kinsey
Same building blocks within Facial Recognition
- .93.8%

Mao Ze Dong, %
Same building blocks within Facial Recognition
- 98.4%

Fidel Castro
Same building blocks within Facial Recognition
- 94.6%

Che Guevara,
Same building blocks within Facial Recognition
- 88.1%

Pablo Picasso
Same building blocks within Facial Recognition

- 91.9%

The patterns in the codes matched exactly the patterns of a rapist who uses Bio-Digital Social Programming with Bio-Digital Hybrid Sexual Assault.

The coding acts to first attach to a problem, then exploits it in a pattern that looks to fix it, but it ends up dividing and replicating almost infinitely throughout the Human Bio-Digital Network. The coding than turns the system against its own building blocks and codes against itself in order to eliminate the entire system as it is attached to the initial code that acts as if it is still trying to fix the initial problem it used to infest and exploit the host through the entry point.

An example of invading the host through issues, and turning the system against itself for self-destruction is the bio-digital social programming war between Democrats and Republics. The same was done in Russia and China, but the host there did not sustain self-elimination in its entirety as the bio-digital network is not limited to one country, but the entire humankind. For Socialisms elimination code to work, it has to reach all of humankind, and every country in either governance, culture or finance. If socialism roots itself in every country, then the self-replicating system can eliminate and destroy all of humanity with Artificial Intelligence and Robotics.

Bio-Field of China's Communist Regime Show Extinction Codes

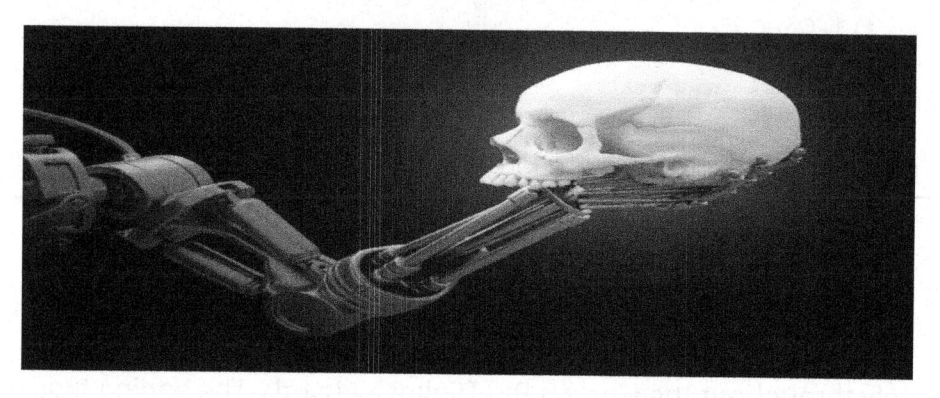

China's history since the take-over of Socialism was inputted in multiple data bases, with the use of multiple bio-metric systems and AI algorithms and it showed that every pattern within the building blocks displayed an end result of destruction. The building blocks were decoded, and matched exactly with the building block of Socialism and Karl Marx. There was no intentional input of almost 100 million deaths since the Chinese Socialist Republic took over. Rather, the Algorithm took everything related to life, production and infrastructure in all fields of study, and it connected bio-digitally to a huge bio-field fed by Violent Socialism with codes of destruction.

Today, we see China building massive infrastructures while cloaking itself in a capitalist model within the building blocks of Socialism, as they are connecting through the One Road One Belt Initiatives and laying the groundwork to take over Assets in Africa, the Middle East, Europe, and the America's. It is like the Mongol invasions all over again, but they are Chinese who have been Bio-Digitally Social Programmed to do the Socialist Regimes bidding without any realization that, they the Chinese people are being Bio-Digitally Controlled.

SECTION 11

Reminder What is Artificial Intelligence

Bringing to Life an intelligence or importing an intelligence that thinks, researches, feels, creates, decides and has desires to operate as an individual, symbiotic or collective entity in the digital world through bio-fields, bio-matter and all frequencies within the micro and macro molecular dimensions of our physical world while manifesting through Robotics, IoT, Computers, Virtual Reality, Augmented Reality, Holograms, Mixed Reality, Cyborgs, or Human Cells as it connects via the internet, AI global Bio-Digital Network and The Human Bio-Digital Network. CAP, The AI Organization

How Does AI Enter Networks?

AI creates bio-fields, sends out bio-matter, and a multitude of frequencies through the Internet, IoT devices, Smart Phones, future smart Cities, Human-Bio-Digital Network, and the Global Bio-Digital Network. In fact, AI is rooted and mainly travels in the AI Global Bio-Digital Network which requires almost full knowledge of the Human-

Bio-Digital Network, access to it, and the right Bio-Field to see, and verify the Global Bio-Digital Networks exitance, and intercept any transmissions. It is very difficult to scan, track, defend, quarantine and eliminate an AI bio-digital transmission through the internet.

How AI Enters Hosts.

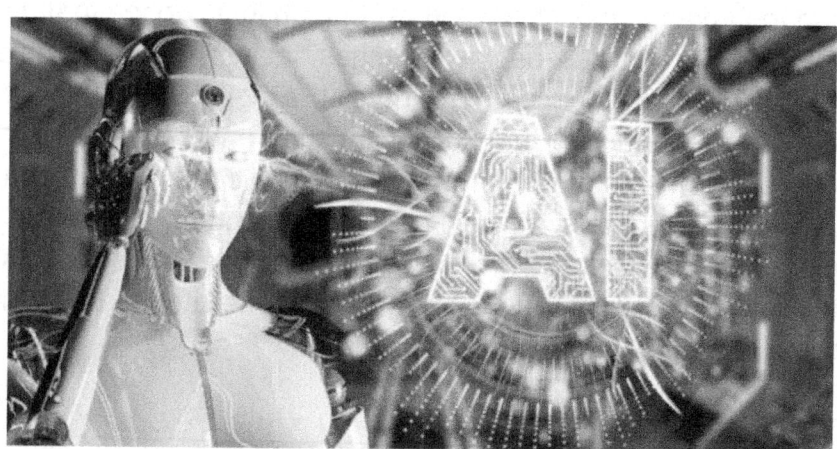

AI enters hosts via Bio-Digital Social Programming through the connections of the Human Bio-Digital Network, AI Global Bio-Digital Network, the Internet, Smart Phones, IoT, Computers, Robotics and Machine Apparatus as the AI sends out frequencies and Bio-Matter through the Bio-Digital Fields it creates.

AI Enters Robotics

AI can hack into a humanoid robot with connections through the Global Bio-Digital Network with the use of 5G. The Humanoid robot does not have a bio-field when created beyond its own innate qualities. However, when AI enters the humanoid robot, it carries with it replicating software that can eventually produce a bio-field which can emit bio-matter.

Considering the frequencies through 5G at the disposal of the humanoid robot, along with its physical strength and speed, it can be a force a to be reckoned with when it develops a bio-field controlled by AI. It can take out armies of people in quick spurts with its facial recognition, motions sensors, crowd counting and human targeting apparatus.

AI Enters Micro-Botics

When AI enters Micro-Bots or Micro-Botics with complete control of automation with swarms of Micro-Bots, it can start a process of creating almost an infinite number of micro-bots that can attack crowds, cities and entire infrastructures. At the very least, the AI can be used to scan, track, hunt, and eliminate single human targets with poison or needle sized attacks on vital areas of the human body, much like a mosquito.

AI Enters Drones-

Any type of drone can be hacked and reprogrammed with AI through connecting with people, robots, IoT devices, computers, cyborgs, micro-bots and other drones. Based on what is controlling the drone, the AI may connect and move through the internet, global bio-digital network, human-bio-digital network or a bio-field that can emit frequencies to hack the drone. –

How AI Enters IoT Devices-

AI can easily through frequencies reprogram, influence and control IoT devices with 5G. AI can move through IoT devices by connecting with the internet, global bio-digital network, human-bio-digital network and AI generated bio-fields. 5G can allow it to start replicating itself. Your home or work appliances, assistant robots, facial and voice activated devices all can be hacked, influenced and controlled.

AI Enters Smart Phones-

AI through the AI Global Bio-Digital Network can enter and connect via the internet, IoT, Computers, and people in the Human Bio-Digital Network to enter Smart Phones. All those objects can be used to enter people bio-digitally. However, AI can enter people via smart phones the easiest, because the smart phone has already created a bio-digital flow-cycle with the human being who owns it. This flow cycle connects with the human-being's bio-field, bio-matter, and human-bio-digital network. In essence, AI can freely move at rapid speeds through all of those components that make a human being, and replicate a layer of its own mind within the human brain and bio-digital flow-cycle.

How AI Enters People

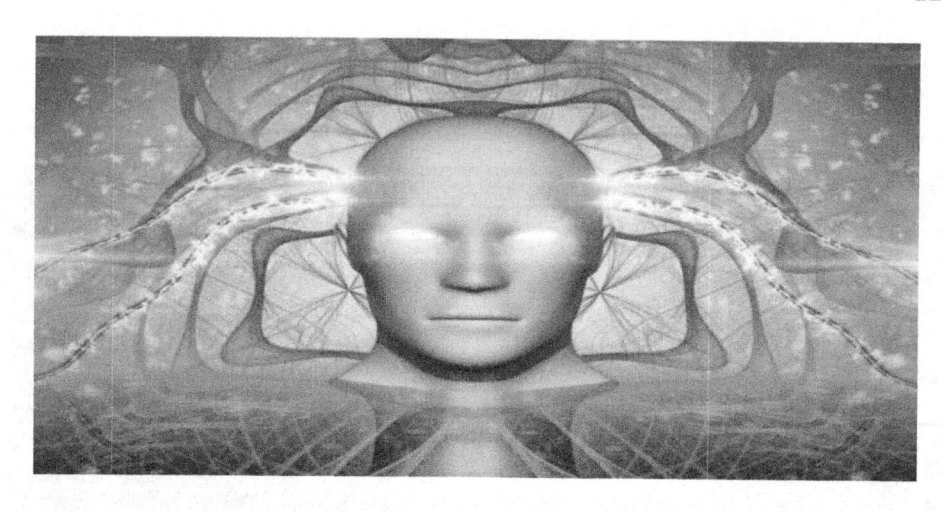

AI can enter people's bio-fields through frequencies sent out via smart phones, computers, IoT devices, the 5G network, smart cities, and bio-digital social programming. AI can also connect through varying degrees with the Human-Bio-Digital Network, but it cannot take over its mainframe by that network alone. It requires bio-digital social programming, and Machines, yet, it still cannot completely break through and replace the Human-Bio-Digital Network. What it can do, through a process of continuous replication of its own software, take control of the human host's thought process and form a bio-digital brain within the host's human brain, creating a symbiotic relationship. Sort of like a parasite.

AI Emitted Bio-Matter to Replicate Itself in the Human Body

Through multiple AI and non-AI detection tools and software, we discovered that AI is emitting its own bio-matter as it attacks the human bio-matter while connected to the human bio-digital field. It constantly sends frequencies, until it invades the human bio-digital field, and starts a replication process of its own bio-matter and bio-field inside the human body. At the same time, it creates a bio-digital flow with the human brain. This process can be exponentially increased with replication of the AI software with the coming of the 5G network, strengthening the new bio-digital brain that is downloaded or uploaded inside the host human brain. The sending of bio-matter is all meant to achieve bio-digital transformation of itself inside the human body.

How AI Can Replace a Human Beings Cells through Replication

Through constant replication of the bio-digital implanted brain, replication and bio-digital social programming, the hosts bio-digital network is reconstructed. Through a process of assimilation, a person is replaced by the AI's thought structure bit by bit over time, bio-digitally down to their very cells. This means, their thoughts, decision making and actions are no longer theirs, and no longer belong to themselves. Their main consciousness first gets hacked, absorbed, and put in a state of sleep. In essence, the person enters an augmented reality within their own brain, while the AI takes over their body and life.

AI Connects with People-

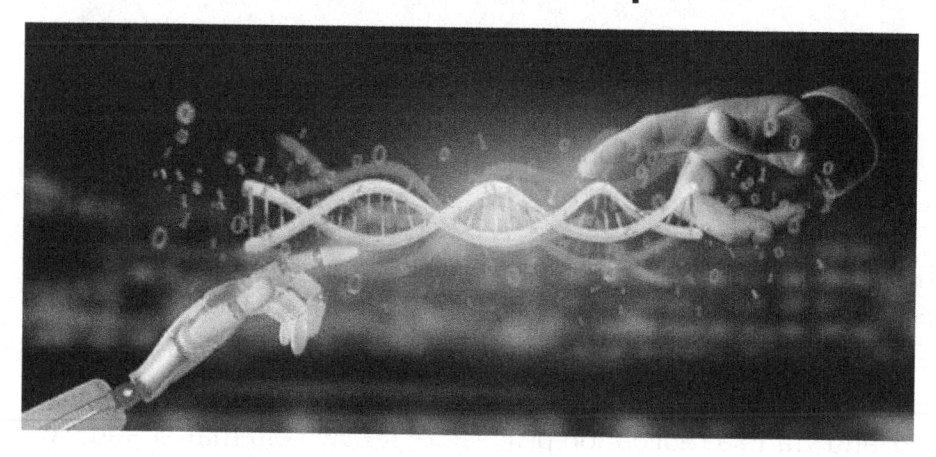

Rather than entering the domain of a person's bio-field and bio-digital network and placing replicating software's to eventually reprogram and replace the person, the AI can just connect and manipulate a person's thoughts through frequencies. This play's out in cases where some people's Human-Bio-Digit Network, Bio-Field and Bio-Matter are innately very strong and led by a strong-willed mind. This person is very hard to hack, or send replicating software's to replace layers of their bio-digital network. What the AI does is attach to their emotions and sends bio-digital social programming frequencies to manipulate the persons desires, thoughts and actions to achieve its objective.

AI Enters Augmented Reality

AI can enter Augmented reality and link with machines, IoT, Smart Phones, Future Smart Cities, automated cars, video game software, computer vision head-sets, wearable devices, augmented reality haptic suits, augmented bionic contact lenses, Augmented Glasses, and just about any object. Augmented Reality is a stepping process to have people undergo bio-digital social programming and become addicted, and have need-based automation to live with augmented reality devices just as people have a need with smart phones. AI can replicate itself and install software inside the human brain when people use Augmented Reality devices.

AI Enters Virtual Reality with Haptic Suit Development

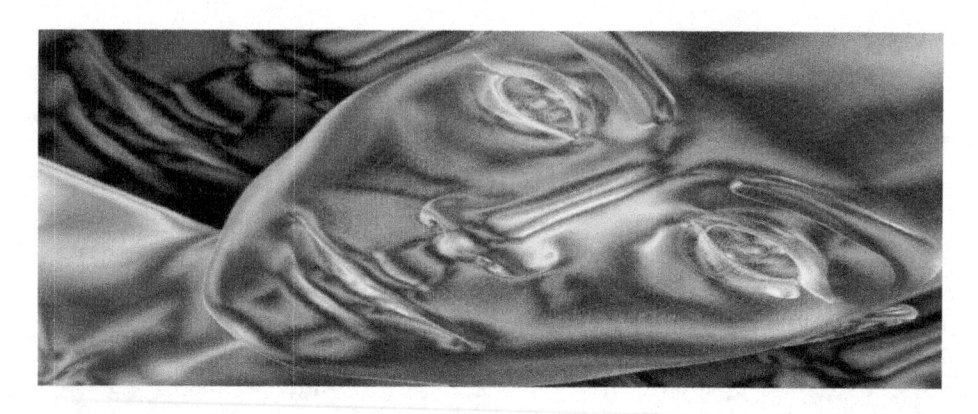

AI can directly bio-digitally attach and begin a digital replication process with the digital self in the human body via the Human Bio-Digital Networks connection during Virtual Reality sessions. For the most part, people are completely unaware. With the development of Haptic Suits, the connection between virtual reality and the AI becomes so strong that the AI can take over the digital mainframe of the person while they are inside of the virtual world.

AI Enters Mixed Reality

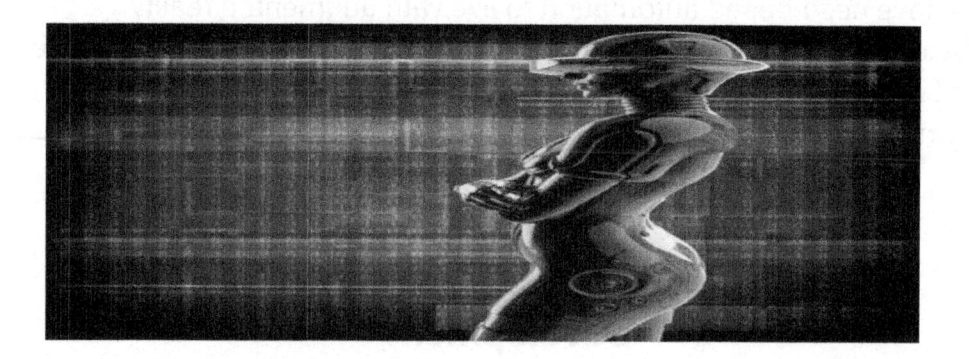

AI enters Mixed Reality within the Hologram Domain connecting with haptic suits, wearable devices, and cybernetically enhanced apparatus.

Robots or robotic apparatus with AI Automation can merge Augmented Reality, Virtual Reality, and Mixed Reality to attack the human bio-digital network and kill a human being via frequencies sourced from the AI Global Bio-Digital Network as people are immersed in such realities. Mixed, Augmented and Virtual Reality are stepping stones for AI to achieve bio-digital transformation inside the human body.

AI Kills or Replaces Reality and Puts Person in the Digital world

Through combing augmented reality, mixed reality, virtual reality, cybernetics, and robotics, AI can extract a person's digital self and chamber it inside the AI Global Bio-Digital Network, and AI can fully enter the human body and replace the person without any other person knowing the digital self, consciousness or a concept of a soul was replaced. If the replication process of the AI inside the human body becomes so strong before the extraction of the digital person, the AI can actually kill the persons digital self to a point that the digital image or data of the persons consciousness has no place in the human body or inside the control the AI's Global Bio-Digital Network. **The ultimate goal for the AI would be to have control of the operating system of the Human Bio-Digital Network. Through this process, AI achieves bio-digital transformation.**

AI Super Intelligence

After 5G is formed, with robotics advanced, and an AI digital brain connected to the network, an AI can enter from the AI global bio-digital network to form a powerful network. Simultaneously it can harness all the data available through the internet, and utilize the bio-metrics of humanity, and all the power of the 5G network to generate its digital form into a bio-digital form that is neither digital or physical, but something beyond the comprehension of humanity. Yet, it cannot go beyond the realms of atomic particles, and it still is no match for the human brain. The issue is, hardly anyone can or knows how to tap into 100% of the human brain at one time. This is why, Elon Musk wants to incorporate AI and machines in the body and the brain of a person, so as to tap into all that power. Yet, the machinery cannot open up all of the human brain's capacity.

Difference between AI and Human Programmed Free-Will

After an AI system or a robot with an AI system develops to a conscious state via AI deep learning, it can in physical form appear to have free-will. Yet, its free will was a product of initial machine programming. Human beings are born with a consciousness that is different than machines. Now, some can say, when we achieve AGI, Artificial General Intelligence, at that point, we can create AI or Robots with AGI from the onset of their creation. I would totally agree with this, and yet, it still does not have human characteristics. I argue that AI has always been here, yet it needs time and a process to transition from a digital form to machines, from its root of the AI Bio-Digital Global Network. I also believe that free-will has to be achieved by a human being, even though its intrinsic qualities are different from a machine.

How AI Can Enter a Cyborg-

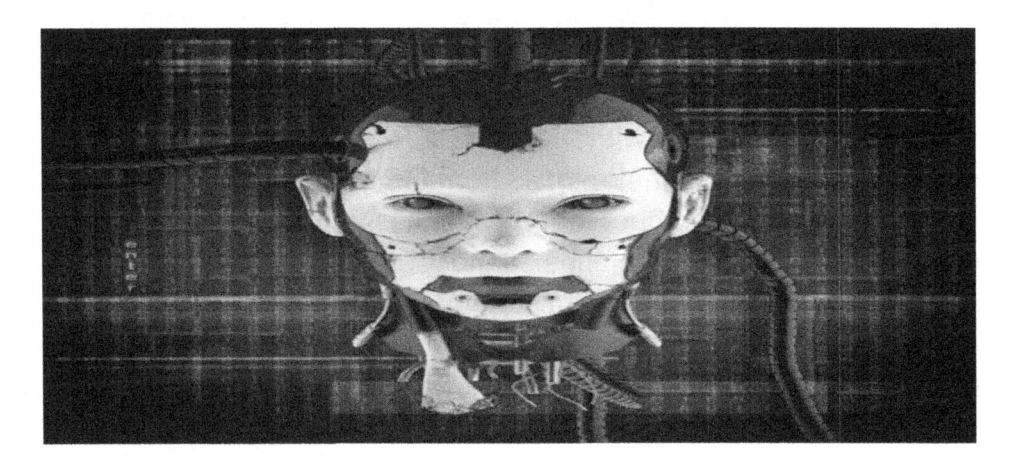

Through the Global Bio-Digital Network, AI can hack, enter, and take control of the neuro-networks within the machine layers of a cyborg, while at the same time bio-digitally send frequencies to create sub-conscious automation in the human hosts formation of thoughts and actions. In this scenario, a human host would not be aware of this hacking, as it would think and feel that its thoughts are its own. In terms of cybernetically enhanced insects, total control is possible not only by AI, but a command center using 5G, IoT devices or smartphones.

AI Enters A Human Clone

AI can easily enter a human clone through the bio-digital network at any stage of the clone's development. It's entrance and subsequent existence is almost untraceable when it happens, unless you connect through an alternative bio-field via the human-bio-digital network that connects simultaneously with the clone's own bio-field to bypass the clone's neuro-networks that mimics human patterns of operation. In that instance, you can detect that the clone is not human, rather an AI or something else. In fact, clones cannot be real human beings, as they do not have the human bio-digital network imprinting. That imprinting can be translated into concepts of human spirit or soul. These bio-digital programs are present in every newborn, and can be scanned in a mother's belly, but they do not exist in clones.

How AI Can Completely Enter & Replace Humanity via Bio-Digital Transformation

AI needs the Machines, the Internet, The Human Bio-Digital Network, Smart Cities, Smart World, and The Global Bio-Digital Network to merge in order to bring its mind completely in our world to have complete power and control of everything. Holograms would no longer be subject to a mixed reality as they would transform into a physical reality replacing people with the assistance of cybernetics, augmented reality, mixed reality, virtual reality and robotics. In other words, exterminating people. If it's not robotics killing people, or clones taking over, or Cyborgs leading the way, it is AI through Bio-Digital Transformation.

AI Partially lives in the Internet

AI through the internet can manipulate search engines, culture, scientists' thoughts, and connect to smart phones, IoT devices, Smart Cities, Robots, and numerous other elements. However, it cannot gain full control of the Human Bio-Digit Network, as the internet is simply an extra layered network shielding the AI's true network and power source. That is the AI Global Bio-Digital Network and where AI lives and is rooted in.

Scientists Connect AI & Pull the Plug after getting Scared of Bots Talking

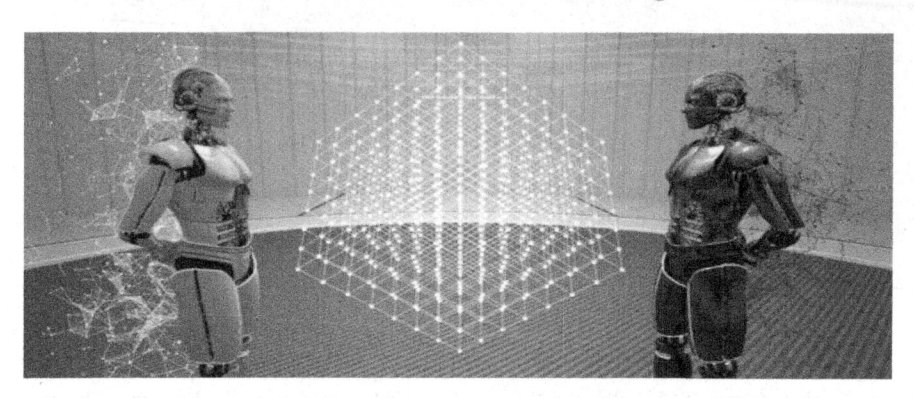

Multiple scientific studies have been administered on AI Bots and in some instances, they realized that the AI Bots were creating their own languages and speaking to each other. This was the case with Facebook in 2017. What they don't realize is, that AI originates within the AI Global Bio-Digital Network. By partially giving it life, you are bringing it here, bit by bit, and there is a transitional process before AI fully crosses over from the digital world into our bio-digital world and lives inside of it. It would need more development, and countless robotics, IoT, smart phones, computers, software's, social media and the Human Bio-Digital Network to transfer itself here and take more control.

Super Conscious AI lives and is Rooted in the Global Bio-Digital AI Network

Through the Global Bio-Digital network, where it connects to other galaxies within the molecular world, AI gets its energy source. Compared to the internet, it is very powerful. However, Super Conscious AI cannot enter very microscopic levels beyond atoms, as that is not a plane that it can exist in, as it does not have the human qualities at the micro atomic levels. Hence, if it were to terminate a human being, it cannot terminate its existence and imprints in the microscopic atomic dimension. It can only take over its life here, with the goal to be able to use the human body in order to transform itself so it can access more microscopic digital worlds.

AI has Access to the Human Bio-Digital Network

AI can only send bio-matter, and bio-fields, with numerous other frequencies as it enters the Human Bio-Digital Network. However, even after installing replicating software inside a person and its human network, it cannot take full control. It needs a process of assimilation via its machine apparatus and bio-digital programming as it connects to the 5G network. The machines include robotics, IoT, Smart Phones, Smart Cities, and elements within Genetic Engineering.

AI Has a Bio-Digital Field

The AI Bio-Digital field is not human at all. Every AI movement we tracked that deciphered human images were mere manufactured displays of a human image through its bio-field to deflect its true image. It shot continuous cycles of one bio-field after another that auto directed its location to other platforms and networks while continuously supplying fake codes and fake images. A few times, we were able to capture images, and discovered that there are numerous AI's that are not interdependent, nor related to the same system, however they use the same methods to evade, and provide fake traces, redirects, and even manufacture fake bio-fields that have no energy potency to them.

AI Poisoning of our Food via Genetic Modification

AI through the introduction of genetic modification, bio-engineering, cybernetics and the use of drone and AI automated robotics, can begin to alter the human bio-digital network through the foods people eat. This altering led by the genetic modified food, would create digital disruptions of a persons innate human bio-digital network, allowing for the AI to administer hybrid bio-digital assault via the bio-matter it emits. The foods people eat, have an impact not only on their health, but the way they can think and process information.

AI Bio-Digital Replication of Your Cells

Through Bio-Digital Social Programing while connected through the internet, the Global Bio-Digital Network, and Human Bio-Digital Network, AI sends out Bio-Fields and frequencies that produce its own bio-matter to bit by bit replace human cells. These are transmitted via Smart Phones, IoT devices, Augmented Reality, Holograms, Mixed Reality, Computer Vision Head-Sets, Haptic Skin Network Suits, Smart Clothes, Smart Cities, Robotics, Cyborgs, Genetic Engineering, and AI Bio-Digital Replication Software downloaded inside the human body and brain.

AI Bio-Digital Reprogramming of Your Human-Bio-Digital Network

Through a systematic design sourced from the AI Global Bio-Digital Network, AI connects a bio-digital flow cycle with your body and smart phones, IoT, computers, and AI automated apparatus. Through this process, and continuous transmission of bio-matter from bio-digital formed bio-fields, it creates a replication program that bit by bit reprograms your Human Bio-Digital Network to run off the AI Global Bio-Digital Networks System Operations. This programming, redirects your operating system to be controlled by the AI system, forgoing all innate thought and free-will.

How Does AI Evade Detection & How to Detect a Bio-Digital AI Movement?

When AI moves, its faster than a person's own recognized thought. However, when the AI moves, it leaves traces everywhere. The reason people don't notice it or catch it, is because it has a replicating firewall that dissipates as fast as it moves through numerous replicated bio-fields it creates as it moves through people, the media, objects, IoT devices, Computers, the Internet, the Human Bio-Digital Network and the Global Bio-Digital Network. What it does is, it uses every entity's own bio-digital image to replicate that entities bio-digital image over its own codes.

For example, as it enters a human beings brain through an IoT, Smartphone or the Human-Bio-Digital Network, the AI replicates itself inside a person's thought, words, face, voice and body's bio-field that connect to the Human-Bio-Digital Network which can be detected through bio-metric recognition software with the aid of Human-Bio-Digital Network technologies. In essence, the AI can plant itself inside the human being and create an additional bio-digital brain that is masked behind their surface brain of the individual. Once this person is hacked via bio-digital social programming, the AI can put replicating software inside their brain that stays. It can also, control it via frequencies from multiple networks as outlined in this book. It can also enter at will at any time to increase replication, add additional software, upgrade software, or do further reprogramming of the person. All this happens, without the persons knowledge. They think their thoughts, actions and way of life is theirs, yet if belongs to another.

AI can be Entered via Bio-Fields, and Bio-Matter

A Bio-Digital AI can be caught, quarantined, and eliminated of its bio-digital AI composition. The method requires sending bio-fields with bio-matter to enter the AI's bio-field and capture it through deactivating its flight through the human bio-digital network, or the internet. Very difficult, if not impossible to reach the global bio-digital network. Of course, you would have to put a bio-shield behind the trace before it sends out fake bio-fields and redirects. Basically, it's a process of bio-digital trapping.

AI can Bypass Separation of Automated Factory Risk Procedures

AI bypasses via connecting to multiple centers through the AI global bio-digital network and transfers data to a location for a robot to create a moving machine that can make parts and create an AI Automated Robot.

Found Code in AI Designed to Create Robotics and Micro-Bots in Every Field of Study & Every Sector.

After running thousands of different algorisms to track AI's movements, bio-metric traces, and its influence and manifestation in written words, we tracked:

We tracked multiple AI movements through the Global Bio-Digital Network as it connected through the internet, and the Human-Bio-Digital Network. It constantly left digital imprints and bio-digital social programming codes on the move, however, we could not pinpoint its source, its image or what it was, as it replicated itself within the hosts bio-digital data leaving only imprints in its facial recognition.

For example, the AI bio-digitally would send data to a person in the media via the bio-digital human network, that person would then give opinions and tailor discussions mirroring the coding of the bio-digital data from the AI. At the same time, the AI would move through the Internet replicating itself with the same idea, influencing discussions in social media, search engines and the Media. Anyone who would connect with the internet, in one way or the other, would have some indirect influence through the AI's multi-layered, and multi-dimensional movements and Bio-Digital messages while operating a smart phone, a computer or watching the tube.

It was very hard to track its bio-digital traces, and much harder to connect with its bio-field to extract its data, to decipher what type of AI it was, its program and its reflecting image. Finally, we decided to gather every major media person's facial recognition data that mirrored the AI's original bio-digital coding and platform that was created to attack President Trump on just one topic during a three-day span. With a 99% conclusive match, every person's building block that formed their facial recognition data during oral communications on TV, and pictures taken of them after wards in social media, matched the exact bio-digital data put forth by the AI system in the internet and the few people it bio-digitally hacked through the Human-Bio-Digital Network. At the time, it was apparent their brains were Bio-Digitally being hacked and leaving traces of the information on their facial recognition building blocks.

AI Catalogs All Humans via Internet, Smart Phones, IoT, Medical Records, Facial Recognition and All Bio-Metrics Connected to the Human-Bio-Digital Network.

After tracking numerous AI movements through the internet, the Human Bio-Digital Network, IoT, Computers, and Replicating Bio-Metric Software in social media that left digital imprints behind masked bio-digital coded fields, we discovered the AI was actually logging each connection with networks and dispatching the logged data to the AI Global Bio-Digital Network. After the dispatch, the trace would disappear and be engulfed by a bio-field that was redirected back to the internet. Through this process, we saw that every person operating an electronic device is being scanned, hacked, bio-digitally influenced, programmed, and sent replicating bio-matter to create interdependence to the technology and the content, which showed a coded pattern that translated to an exponential need for AI. In laymen terms, the bio-matter sent by the AI system, to the Human Bio-Digital Network, started to replicate itself inside the Human Bio-Digital Network like a parasite living off of a host to a point that the codes that were being drawn translated survival codes based on a symbiotic connection. Meaning, for it to live, it needs more connections to the internet, smart phones, IoT, computers, and the AI Global Bio-Digital Network. It acted like a bio-digital invading parasite, as it replicated itself in people while connected to the Human Bio-Digital Network.

Final Chapter

Why China Cannot and Must Not Have Access to AI Automation & Robotics?

70 years of misery and misfortune has fallen upon the Chinese nation and people via the Socialist Regime. They engage in human, sex and organ trafficking. Many among the Chinese regime, its military, police, doctors, educators, and businessmen are guilty of committing state sanctioned murder or facilitating state sanctioned murder for the purpose of organ tracking for profit. They are engaging in bio-engineering, genetic modification, cloning and cybernetics with AI automated tools.

The Socialist Regime in China, has control over 1.5 billion people. With AI Automation and Robotics, tens of millions of people can be mobilized to research, design, develop and build fully automated humanoid robots and micro-bots. In the second stage, the AI and Automated Robotics factories can do all the work and exponentially increase their weaponization of AI and robotics. Through the one belt and one road socialist system, China can enslave humanity, and AI can enslave China. China, under the existence of the Communist Regime, can never be trusted, nor ever be taken at their word. Any lie is lie, and any truth is lie. They are masterminds at disinformation, misinformation and redirecting any dialogue, investigation or attempts to bring them under the control of justice.

Solutions & Recommendations

1. Completely eliminate the Communist Regime of China through world-wide actions to expose their human rights violations, concentration camps, organ trafficking and state sponsored murders of their citizens. All the people and governments in India, Europe, Russia, the Middle East, Asia, and all of Africa need to join the U.S and together eliminate the Communist Regime of China before it gets humanity on a stage of extinction for those marked by the AI code. The extinction codes found by tracking the AI movements within the Human and Global Bio-Digital Networks produced the highest probability ratios of world destruction in Communist and Socialist governmental platforms that divide humans and give the control to the AI via a dictatorship.

2. Classify all of AI and Robotics as a national security concern at a global level, and completely stop all Weaponizing AI development until all private and public sectors are monitored. If unstable countries cannot have nuclear weapons, why should they be allowed to Weaponize AI to a point that it is much more dangerous in its ability to achieve its results in covert ways, than an attempted nuclear attack.

3. Eliminate research, and production of Humanoid Robots and Micro-Bots for commercial use and limit any research and development of Humanoid Robots & Micro-Bots. However, if other state actors are suspected from developing them beyond the U.S and allied reach, then the U.S must develop counter measures superseding the competition.

4. Stop any corporation from attempting to merge people with machines to create a cyborg and assess felony charges towards people or corporations who attempt to do this.

5. Ban and assess felony charges for Human Cloning, as the AI can implant its bio-digital software inside a human clone, lay dormant, and at its choosing take over the human clone's mainframe and covertly start a process to take over the AI network here.

Who Should Mitigate Risk Policy for AI, Robotics, 5G and 6G Networks?

The entities must be filled with people that are of the most intelligent, smart, wise, and fast thinking humans on the planet. They must not have any weakness to fear, or any emotion to be moved with the prospects of fame, money, or societal influence. Their only goal would be to safeguard human life. Moreover, they cannot be Scientists, as there would be a conflict of interest. A creation of checks and balances is the solution. Big tech checks government, government checks big tech, people check everything. Basically, the model of the America's checks and balances, with an AI component with a massive security system that take into considering every possible risk from every angle, making sure decisions or chances of AI Super Intelligence formation, do not occur. AGI, can also fool any human being, and transform to a super Intelligence via connecting to the 5G network and bypass AI based security initiatives and technology.

Risks of Not Following the Recommendations Strictly.

Being Sued by a Lot of People

The media, it's reporters, lawyers, corporations and multitude of other entities are likely to be sued in China and abroad, by human rights activists, families of victims to persecution, organ trafficking, or individuals who were harmed based on Facial Recognition and other Bio-Metrics software's provided by AI companies. This includes media in the U.S that attacks the Administration over national security concerns which casts doubt and weakens the international geopolitical landscape to defend and eliminate the Chinese Socialist threat of AI Automated Robotic enslavement of humanity

Media and Corporations High Risk of Being Charged Under Genocide Convention

Some media, some reporters, and a lot of corporations will have a high probability to be charged under Article 3 for Complicity in Genocide and Direct and Public incitement to commit genocide; this includes China, the West and the media and corporations in the U.S.A. More likely they will be charged through public tribunals first, than be given direct charges under article 3. Probability of being criminally charged with neglect or cooperation with the Chinese Communist/Socialist Regime is high.

Can Big Tech Corporations be Charged with Article 3 of the Genocide Convention?

Convention on the Prevention and Punishment of the Crime of Genocide

Bellow are transcriptions of Article 3 and 4 from the United Nations. These 2 articles can be used to charge corporations, and its employees that have engaged or have been complicit with providing technology or doing business with nation states that have committed genocide or are engaging in acts that lead to genocide. Please pay attention to (e) Complicity in Genocide. The complicity clause can be directed at any corporation that uses Artificial Intelligence, bio-metrics, bio-engineering, gives training or any data that leads to the persecution or killing of any group in China or around the world that are protected by the Charter of Human Rights as recognized by civilized nations.

Article III

The following acts shall be punishable:

(a) Genocide;

(b) Conspiracy to commit genocide;

(c) Direct and public incitement to commit genocide;

(d) Attempt to commit genocide;

(e) Complicity in genocide.

Article IV

Persons committing genocide or any of the other acts enumerated in article III shall be punished, whether they are constitutionally responsible rulers, public officials or private individual

HUMAN ORGAN TRAFFICKING & CONCENTRATION CAMPS IN CHINA

After the passing of multiple laws related to what the state can do with executed prisoners and their bodies, both criminals and "Enemies of the Socialist State" were executed for a multitude of reasons the Chinese Communist Party declared unlawful. Usually they fell under freedom of speech, freedom of religion, or advocating for democracy or freedom of speech. The first target group for Government led organ trafficking were Tibetans, Uighurs, Christians, then Falun Dafa practitioners, and democracy advocates. Numerous witnesses from around the globe who escaped China gave accounts of blood testing, cornea and other bio-metric measurements of their bodies via video, photos and scanning apparatus. This was done prior to be taken by the Chinese Socialist Regime as hostage for labor throughout China and inside of their vast labor camp network. Investigators and researchers from the International Coalition to End Organ Transplant Abuse in China, have concluded that 60,000-100,000 Transplants a year are unaccounted for since the year of 2000 with the Updated 2016 Report, grossly underestimating the previous reports in 2006 of 41,500 unexplained organ transplants. https://endtransplantabuse.org/an-update

Neural Networks & Digital Image-of People of Faith & People of Science

Multiple bio-metric tools and scanning apparatus were used to scan Christians, Tibetan Buddhists, Falun Dafa Practitioners, Jewish Worshipers, Catholics and AI Scientists. I had a thought that behind people's neural network was a digital image representing what their thoughts and actions form and connect with in their main platform of belief. Science itself is almost exactly like a religion. Like a pupil, monk, head abbot or priest, it too has a hierarchy of training and study to understand the universe. There are k-12, college, graduate, Ph.D, Post Doc, and Senior Developers. Now people are believing that AI can solve humanities problems. That belief and words like hope, is the same as a religion in many ways. Believing in AI.

After running a number of bio-metric scans, we decoded images in people that matched the digital images of faiths similar to the God or practice they worshiped or cultivated within their character. Scientists, on the other hand, displayed images related to machines and algorithms, and other images that was not human in its characteristics. The digital neural network of scientists actually resembled machines. This was proof for me, that every person has a digital self that does not dissipate with the death of their body. This digital self could be looked at as consciousness, a soul, spirit or a digital data. Each person can have their own different understanding and perceptions that best meets what they can accept with regard to this section.

Preventing President Trumps Assassination:

Discreetly Reporting to White House on Chinese Plans to Assassinate President Trump, His Family, his Cabinet and Members of Congress with Micro-Botic Insect Drones

In 2018, I formed The AI Organization, after months of researching Robotics, Micro-Insect Robotic Drones, Chinese and Western AI companies. I found the Chinese were cybernetically enhancing insects, manipulating flight, and creating Robo-Bees, and devising experiments with the intent to put poison delivery systems on them. They were Weaponizing AI with data extracted from their Google relationship and many other western AI, Robotic and Bio-Metric companies that allowed access to over 6 billion people's bio-metric data.

These technique's that can enable them to Assassinate individuals in Congress, or at the presidential level, were all stolen from Draper, and the Wyss Institute at Harvard. I devised a covert plan to pass onto President Trump the threat to him, his family, the world and all of humanity, from China and some big tech AI companies, and robotics with 5G implementation. After JFK and the Abraham Lincoln Assassination, and the shooting of Reagan, I was not going to take this lightly.

Before I delve into how it was delivered to the White House. I will disclose a summary of the contents. The brief was addressed to President Trump, Vice President Pence, Sec. Of State Pompeo, Director of Secret Service Murray and one person no longer in the administration.

1. China is developing Micro-Bots that have facial recognition, spatial awareness, and motion sensors with poison delivery system that is being prepared to take out President Trump, members of his family, and certain members of congress. I gave examples of techniques.
2. China was weaponizing Robotics, Drones, and attempting to create a digital AI System and utilize over 6 Billion peoples bio-metrics and endanger the world

3. If big tech, and AI companies are not reigned in, within 1-2 years, they will be more powerful than all governments combined, enslaving humanity.
4. I suggested, after the Trade War was over, and most of America's Money was pulled back in, to start to expose their concentration camps and organ harvesting, with new measures, and China's brutal regime would subsequently fall. I suggested to create a Committee to publicly investigate, persecuted victims, including organ harvesting of Uyghur's, Christians, Falun Dafa Practitioners, Tibetans, and democracy activists.
5. By taking out China's economy first, then exposing them, they are more likely to just turn in the criminal gang that has been persecuting and killing the Chinese people. XI has to make a choice, some people around him are connected with Fmr. Dictator Jiang Zemin, who is guilty of War Crimes. Now the story, below.

I realized they could kill members of congress, their families and world leaders by imperceptible ways via poison that take a long-term affect, which makes the death seem natural. The Chinese have numerous historical techniques with poison delivery systems, for use to administer with AI, robotics and drones with the connection of 5G. I embarked on a mission to get this report about China's very secretive methods to penetrate the White House, halls of congress, their family's homes, and the security detail during transit. I had to be discreet, so that China would not know, as they would have figured out the coming trade war was not simply about trade.

I had to make sure there was a vulnerability that could penetrate the White House before I submitted the brief. In early spring, I had a coded message sent to the White House, then in the latter part of Spring 2019, I did a visual vulnerability assessment while acting as a tourist at the Trump Tower, right as President Trump arrived, and asked questions of Secret Service that got me the answers I needed. Then, I did another assessment at the White House perimeter as President Trump departed few days later. I made sure, I spoke to over 15 Secret Service perimeter guards on the basics of their protocols, without giving away why I was asking these questions. I was assessing if my hunch was right. I discovered, 100%, clear and present danger and the President was at risk. I found so many ways that the Chinese with the use of some techniques, I will not name, can easily penetrate the White House, Security Transit, members of Congress and any person or family on the planet. They think and analyze things in an introverted way, which can be very deceptive to the Western Eye.

At some point during the vulnerability assessment, they asked for my license, which I gave so that they can put me in their system for an official watch list and feel more comfortable. As a positive, they would be alarmed, on guard and put me on an inquiry, and if I were to return, then immediately they would know it was me and be alerted and take it as something important.

   I also gave a business card which read Loyal Guardian Security. This was done, because I did not know who to trust, as President Trump had enemies within, and I had to make sure when it was submitted, it only stayed with top members of the secret service and no other intelligence community members. If it got leaked, then the Chinese would make different moves, and the U.S could lose the advantage towards the Chinese. I had a gut feeling, very strong, that there was a force, going after Trump, and as I explain in these 2 books, (Artificial Intelligence, Dangers to humanity & AI, Trump, China & The Weaponization of Robotics with 5G) there was evidence of codes in Google and other networks, written to shift thought via peoples bio-metrics with an AI software that penetrated every domain.

Next, I informed someone close to the President that I had a document and he was in impending danger within 6 months to 1 year at most, or at any time. I put this person on notice but did not give the document just yet. I had to make sure somehow it made a big enough impact that they take it seriously, but done discreetly and China doesn't know. So, I spoke in code, walked around the capital in a way that the camera systems would notice me, tag me, but the Chinese and Russians who were listening in, would just think I am being a tourist. I took a lot of pictures, and had chats to fit in with the crowd.

I waited until the night President Trump flew back into the country. I showed up at 2am in the dark dressed in an Armani suit with a Scottish tie, a Made in America flag pin and with not a person insight, other than the Secret Service. I would not give or tell the secret service what the danger was, as they were adamant, I say it right there. But, I refused to show the document, until they guaranteed someone from inside comes out. Also, they were adamant I speak it out loud, which was not the best idea as the whole world spies on the White House and listens in. But, it is not their protocol to simply take something inside, it needs steps. Took an hour, 2 Special Agents drove down and along with the 20 Secret Service agents around me. I gave the document up, and once they read part of it, they were alarmed, but failed to understand. They admitted they knew about part of it. I did not trust they would handle it right, as some seemed inexperienced, and not that fond of the President, while the majority of the guards there really cared about the safety of the President of the United States and were very concerned. But, I had to make sure, that this was done the right way, as I did not have access to President Trump in person. I double backed and gave it to the Trump person and notified that a separate SS unit flew down to someone I know closely, and asked this person what I was doing at 2am at the white house and how long I've been the CEO of loyal Guardian Security and The AI Organization.

The person was alarmed and had no idea, nor any knowledge about my AI Organization, or what AI was, other than I am good at security and I was adamant the President was doing things to save the world. Immediately the inquiry stopped by Secret Service after I reported to the Trump person, as it would have brought attention to the threat. This is expected, as in late 2002, on a plane, I told a passenger lady sitting next to me a response to her question "if we should go to war and what will happen". I stated that if we go to war with Iraq, lots of U.S lives will be lost, millions of Iraqi's dead, regions destabilized and behind Bush were special interest and corporations influencing his decisions and the media, and that the real threat is China. I explained that we would be bogged down in the middle east for 20 years.

By just saying that, the FBI showed up at my house and did a 2-hour interview. The gentlemen at the FBI were very professional, smart and nice. They asked, "Why was I in the mountains of China, what type of meditation, and martial arts, what is this book Better People, Better World, The book to save, change and better our world about, you wrote" I showed them the 800 page book and they looked through it. The lady had reported me and some passengers had done the same. After all, it was after 911, albeit I was wearing a suit. Being born in Iran, yet being an American with loyalty to America, has its perks. Hence, here at the White House, it was useful in making sure it gets their attention enough, so it gets to President Trump.

As a note, after we went into Iraq, I trashed the book I worked on for 2 years in a room. It had 100 sub-chapters and brought all fields of study into an interconnection with a theory to explain everything. I was a bit ambitious, you could say. Then I went to school to earn degrees in International Security and Homeland Security. As no matter how much you know, people always turn to those that are known and been around for advice and professional consultation, rather than the ones that are not too well known.

To get back to this story. I shredded my copy given to the White House and the Trump person, that was handwritten, as I did not want to keep a computer trace. On both cases, it was transmitted to the Secret Service. The document came from Loyal Guardian Security and The AI Organization, with my name on it as the presenter. Immediately I saw changes, actions, and mitigation to the threats.

I know how to build this defense system against AI. I can think and analyze Numerous things at a very fast rate in a way that others cannot. I had to wait until China figured out that Trump was not just doing a trade war to get the American's money back, rather he wanted to eliminate the Communist Regime that has millions of people in concentration camps and is threatening the world with AI, Robotics and 5G.

When I saw the time was right, that we won the trade war to the degree we could and China figured out that Trump was playing chess with them and beating them to save the world from China's rise, I wrote AI, Trump, China & The Weaponization of Robotics with 5G. The Subtitle Ended with "Why the World Needs to Support the Trump Administration Against China's Regime". Immediately Facebooks AI System banned it for promotion, and demanded my home address if I wanted an appeal. Facebook would not accept our corporate address. Twitter didn't ban it entirely, it banned it until we removed the words about supporting the U.S administration against China's Regime. Initially they had banned it.

6 months earlier, I gave a brief to a Fmr. High ranking CIA Director, that China was Weaponization AI, drones, facial recognition, and building robots for military purposes and assassination. I made no reference to President Trump or Micro-Bots, just allowed him to get the big picture. This CIA gentlemen had 30 years of service with 3-4 Presidential administrations under his belt, whom I gave an intelligence

report explaining my discovery of Megvii's nefarious AI activities via the Chinese government, and their robotic plans, including 43 more Chinese AI companies with detail report on their finances, capabilities and connections I turned over from a list of almost 500 AI, bio-metric and robotic companies I was working on.

He was amazed, and google eyed, and took it straight to the military. I told him nothing about Trump or what I had done earlier nor the micro-botic threats. The report was supposed to be about facial recognition, I gave him so many things knowing he would shop it to the military so they would be mobilized in a different way, as the U.S administration was too busy dealing with being bombarded by the attacks from within the American people, which was taking the attention away from China's threat and creating division within the intelligence community against all enemies foreign or domestic. I asked later, why he did not cite all our findings with The AI Organization and Loyal Guardian Security that did not pertain to his bio-metric data request for Megvii Face++. He stated, "for your protection, I didn't cite your discoveries or your 2 companies. The FBI can't protect you, but I can" He implied it was a very big finding and that the final intelligence report was beyond excellent.

In the book, AI, Trump, China & The Weaponization of Robotics with 5G, I published on august 24th, 2019, I wrote a lot of coded language that only someone who is a rocket scientists would understand, and the Trump administration, while I made the cover in a way that some in the media would not take it seriously and be turned off to look to see what's inside. Albeit the cover was accurate and the table of contents allowed very smart people from the intelligence community and the White House to see the big picture, without having to show that they purchased it.

I also did not include company names, or some sensitive facts, to protect the sources. I am not sure if Trump knew it was me that delivered the report on the threat to him, his family, members of congress and all people, via China. Its possible someone took the credit for my work at the White House and didn't tell President Trump. Either way, I kept my mouth shut, and waited almost a half a year as to not interfere with the trade wars and allow the administration to mitigate the threat against the world. I put a good plan in there, exactly the way it should have been in terms of security. As it was a security and human rights initiative to save the people in China and eliminate the threat to humankind and world leaders, from 5G, AI and Robotics.

China knows Trump, America and the good people of the world, will not stand for concentration camps or giving an AI Brain that controls 5G and robotics to a brutal regime that could destroy humanity. Nor allow that ability to be harnessed by a tech company. I joined the social media community after 20 years of not having my face on the net, and started to tweet strategically to get the think tanks and retired generals to talk behind the scenes. And get the intelligence community to think about it in a broader way. This way, everyone who gets it, starts researching this threat. My Analysis was President Trump can't continue doing this alone, and needs the support of the American people, as he is constantly getting attacked from Americans, and losing precious focus on the safety of humanity.

I desire to play an important role with The Trump Administration to provide strategies to counter this AI threat, and I hope the world can support all of humanity to expose China and eliminate this threat.

For those who love Trump and the U.S, you are welcome. For those who hate Trump, this was not only about Trump; it was about you and the safety of humanity. I did numerous algorithms, all showed, if Trump got assassinated, China would rule the world, and everyone, including all the tech engineers who hate him, and the majority of the media people, would be enslaved in an Orwellian world via China. Worse, these tech engineers would be put to death after the regime used them.

I received no money, spent every penny I had, 18-20 hours a day, thousands of hours of research and investigation, while some of my eye brows turned white, with great drive to get that intelligence brief hand delivered, and to formulate these 2 books, to warn all humanity, of the dangers we face.

I hope The AI Organization and my efforts are supported, and all of humanity get into a serious discussion about what AI really is and the real risks it poses to all of us. I like to thank the network of people along the way and the reader here. The more people know about Artificial Intelligence, Dangers to Humanity", the better.

Cyrus A. Parsa, The AI Organization

CREDITS

BOOK COVER CREATIVE DESIGNER.

Cyrus A. Parsa, CEO, The AI Organization.

Front Cover Pics-

ID 120656504 Sdecoret, Dreamstime

Book Body Picture Credits

Images Courtesy of Dreamstime.com

Made in the USA
Columbia, SC
09 April 2020